Pictorial Atlas of Soilborne Fungal Plant Pathogens and Diseases

Pictorial Atlas of Soilborne Fungal Plant Pathogens and Diseases

Tsuneo Watanabe

CRC Press
Taylor & Francis Group
Boca Raton London New York

CRC Press is an imprint of the
Taylor & Francis Group, an **informa** business

CRC Press
Taylor & Francis Group
6000 Broken Sound Parkway NW, Suite 300
Boca Raton, FL 33487-2742

First issued in paperback 2021

© 2018 by Taylor & Francis Group, LLC
CRC Press is an imprint of Taylor & Francis Group, an Informa business

No claim to original U.S. Government works

ISBN-13: 978-1-03-209578-3 (pbk)
ISBN-13: 978-1-138-29459-2 (hbk)

Library of Congress Cataloging–in–Publication Data

Names: Watanabe, Tsuneo, 1937- author.
Title: Pictorial atlas of soilborne fungal plant pathogens and diseases /
author: Tsuneo Watanabe.
Other titles: Mycology series ; [v. 33]
Description: Boca Raton, FL : CRC Press, Taylor & Francis Group, 2018. |
Series: Mycology series ; [33] | Includes bibliographical references and
index.
Identifiers: LCCN 2017031969 | ISBN 9781138294592 (hardback : alk. paper)
Subjects: LCSH: Soil fungi— Atlases. | Fungal diseases of plants—Atlases.
Classification: LCC QR111 .W2675 2018 | DDC 632/.4—dc23
LC record available at https://lccn.loc.gov/2017031969

Visit the Taylor & Francis Website at
http://www.taylorandfrancis.com

and the CRC Press Website at
http://www.crcpress.com

Contents

Preface

On the science related to soilborne fungal diseases of plants and pathogens, there are numerous aspects and fields as part of the environmental sciences; thus, some of my own published works were selected as introductory topics, because they are basic, and because they may engage audiences by showing hosts damaged or the diseases with more emphasis on the contrary to fungi themselves in my previous work (Watanabe, 2010). Each topic must be readily accessible through illustrations, together with simple explanations.

I hope this will be the first step for any audience with further interests and for future studies.

I would like to thank John Sulzycki, senior editor; Jennifer Blaise, editorial assistant; Marsha Hecht, project editor; Fiona Macdonald, publisher, CRC Press, Taylor & Francis Group, LLC; and Linda Alila, Associate Project Manager, diacriTech, for their review of the manuscript after John's retirement at the end of July.

Tsuneo Watanabe, PhD
Tsukuba, Ibaraki, Japan
September 16, 2017

Acknowledgments

The following journals are acknowledged for granting copyright permissions: *Annals of the Phytopathological Society of Japan*, and *Journal of General Plant Pathology* (the Phytopathological Society of Japan); *Mycologia* (the Mycological Society of America and the New York Botanical Garden); *Mycologia Helvetica* (the Swiss Mycological Society); *Phytopathology* (the American Phytopathological Society); *Transactions of the British Mycological Society* and *Mycological Research* (the British Mycological Society); and *Transactions of the Mycological Society of Japan* and *Mycoscience* (the Mycological Society of Japan) in relation to the previous works.

Author

Tsuneo Watanabe received his BA from the University of Tokyo in 1961 and his MS and PhD degrees in plant pathology from the University of California, Berkeley, in 1964 and 1967. Dr. Watanabe worked as a plant pathologist and mycologist in the field of soilborne diseases of plants, their pathogens, and soil fungi at the National Institute of Agro-Environmental Sciences from 1967 to 1983; at the Forestry and Forest Products Research Institute from 1983 to 1997; and after retirement, as an invited research scientist at the National Institute of Advanced Industrial Science and Technology (AIST) at Tsukuba from 1997 to 2015 including a five-year engagement in the national projects. Over the course of his career, Dr. Watanabe has published approximately 170 scientific papers on the subject and three books titled *Photomicrographs and Illustrations of Soil Fungi* (Soft Science Publications, Ltd., Tokyo, Japan, 1993), *Dictionary of Soilborne Plant Diseases* (Asakura Publishing Company, Ltd., Tokyo, 1998), and *Pictorial Atlas of Soil and Seed Fungi* (1st, 2nd, and 3rd eds., CRC Press, Boca Raton, FL, 2010).

Manual Design and Usage

Design and usage of this book are summarized as follows:

The fungus diseases and topics related by various plant pathogenic species are alphabetically described, including oomycetous, zygomycetous, ascomycetous, basidiomycetous, and deuteromycetous (anamorphic or mitosporic fungal) species.

Latin binominals are adopted, following recent literature, including Farr et al. (1989), Jong et al. (1996), Kirk et al. (2008), and NITE Biological Resource and Center (NBRC, 2005).

Most of the fungi studied were isolated from soil, plant roots, and seeds, and the rest from wood-inhabiting fruiting bodies. Their spores or the spore-like structures associated with them were mostly collected in Japan, but some were from the Dominican Republic, Paraguay, Switzerland, and Taiwan, the Republic of China (ROC).

Living fungal cultures are deposited at the MAFF Genebank, National Institute of Agrobiological Sciences, Ministry of Agriculture, Forestry and Fishery (MAFF), and the Institute of Biological Resources and Functions, National Institute of Advanced Industrial Science and Technology (AIST), both in Tsukuba, Ibaraki, Japan, as well as at the Biological Resource Center, National Institute of Technology and Evaluation (NITE) (previously Fermentation Institute [IFO], Osaka incorporated) in Kazusa, Chiba, Japan, at American Type Culture Collection (ATCC), United States, and at Centraalbureau voor Schimmelcultures (CBS), Netherlands, and they are mostly listed in the *Appendix of Pictorial Atlas of Soil and Seed Fungi*, 3rd ed. (2010). Most fungi have been deposited domestically, but foreign GenBank accession numbers have been recorded preferably.

All pictures on morphologies, and fresh and dried fungal specimens are described based on my own materials, and reproduced or created using Adobe Photoshop Elements 5.0. Dimensions of some representative organs of the respective fungi are also recorded whenever necessary.

Dried specimens were prepared: fruit bodies dried at 50°C until completely dry, stocked one by one in specific plastic cases; similarly, their fungal agar cultures in plastic Petri dishes dried at 50°C until completely dry, then removed from the plates, and filed in albums one by one. These specimens were always kept nearby and used.

Dried Fungal Specimens

Specimens are most significant for research, for establishing continuity beyond time and sites, and are basic and essential for any works. Although live specimens have been preserved in the public culture centers (that is, ATCC), most dried specimens have been neglected. These specimens were preserved simply around us.

DRIED FUNGAL SPECIMENS

1. Two *Pleurotus ostreatus* isolates, each in test tubes and in plastic petri dishes.
2. Dried cultures filed together.
3. Fruit-body-forming specimens stocked in slide cases.
4. Dried fruit bodies: *Armillaria, Morchella*, and an unidentified basidiomycete from left to right.
5. Dried culture specimens: *Pythium, Mortierella, Mucor,* and *Alternaria* (top row, left to right); *Nectria, Sordaria, Flamulina,* and *Armillaria* (bottom row, left to right).

The Contents Covered

ETIOLOGY

Studies related to *etiology*, or the discovery of causal agents, are the most important investigations of the soilborne fungal disease problems of plants and should be the first approach to any problem. The procedures have been taken to purely isolate the pathogenic organisms from diseased plants, soil, substrates, or air. Then, they are separated from one another, and subsequently, the respective organisms are tested one by one for the pathogenicity with the most suitable inoculation methods under the most appropriate environments. Koch's postulates must be proved during the process of etiological works.

It is possible to know the respective organisms in relation to the host plants and its pathogenicity. In isolation procedures, media and culture conditions, including temperature, humidity, light and pH, are selected or tested. Furthermore, inoculation methods must be considered in each procedure or toward particular organs. The organisms to inoculate are to be selected and tested directly on the host plants, through seed or soil infestation.

IDENTIFICATION, CLASSIFICATION, AND PHYSIOLOGY

The pathogenic organisms are possibly identified and classified molecularly, morphologically, and physiologically. Once experienced, the organisms must be identified readily only by watching the morphologies. However, the taxonomical problems have been solved with satisfaction in all of these applied approaches. The physiology of the fungi must be significant for the pathogenicity and may include the optimum temperatures for growth, sporulation, or reproduction of each fungus.

ECOLOGY

The fungi may be studied ecologically. How does the fungus survive? Where, when and/or how does it start to germinate, contact with the host, and penetrate the plant? The survival abilities are important to know, including morphologies of the resting structures in soil or any environments without any host plants.

DISEASE CONTROL

The contents covered in this book are essential for knowing the diseases that fungi cause, ways to protect against diseases using biocontrol or ecological measures, and how to achieve healthy environmental conditions without traditionally used chemicals.

1 *Aphanomyces* and *Plectospira*

The genus *Aphanomyces* is characterized by forming zoospores, which move in a row inside hypha-like sporangia and encyst in a mass at the tips. Oogonia are terminal, often aggregated, include aplerotic oospores, and bear monoclinous antheridia suppressed often on coiled hypha-like antheridiumphores. Plant-pathogenic *Aphanomyces* species have been isolated from tissues of various host plants worldwide (Drechsler, 1929). *A. cladogamous* was isolated from tomato field soil or forest soil in Japan by bait methods with cucumber or lupinus seed.

The genus *Plectospira* is morphologically similar to the genus *Aphanomyces*. It forms lobate sporangia connected with hypha-like sporangia; and its zoospores, after moving in a row inside sporangia, encyst in a mass at the tips just like *Aphanomyces*. Oogonia is surrounded by monoclinous and diclinous hypha-like antheridia, including both plerotic and aplerotic oospores. Oospores are often formed without sexual contact of both oogonia and antheridia parthenogenetically.

Plectospira myriandra (isolate 84-209 (ATCC64139)) was first reported as tomato root isolates in the United States (Drechsler, 1927) and re-isolated in bamboo fields in Tosa, Japan 59 years after its discovery by the cucumber seed bait method (Watanabe, 1987); however, this fungus has now been commonly isolated from various areas in Brazil (Pires-Zottarelli, 2011) since its first record of isolation in Brazil (Gomez et al., 2003).

A. Damaged radishes harvested from a severely infected field in Akita.

B. *Aphanomyces* sp. associated with the underground disorder of radishes.

(A)

(B)

1.1 MORPHOLOGIES: *APHANOMYCES*

Aphanomyces cladogamous (isolate 83–413) (**A–E**)

A. Hypha-like sporangium bearing encysted zoospores at the tip.

B. Zoospores moving inside the exit tube, and two zoospores encysted outside.

C–D. Antheridia and oogonia bearing oospores.

E. Winding antheridiumphores, and antheridia surrounding oogonium, including single aplerotic oospore.

(Zoospores encysted 7–8 µm in diameter; oogonia 18.7–35 µm in diameter; oospores 15–25 µm in diameter)

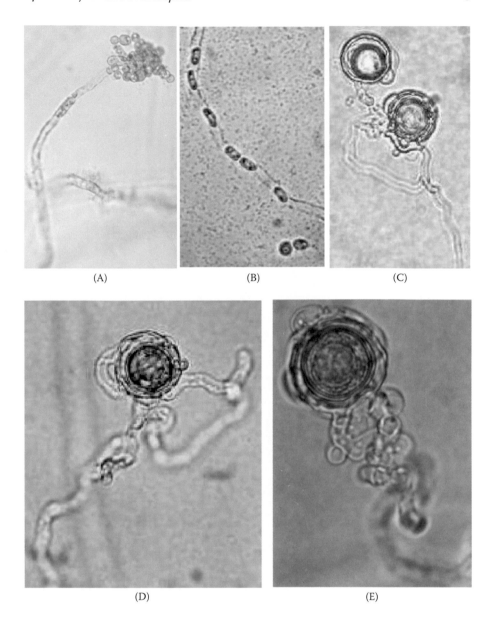

(A) (B) (C)

(D) (E)

1.2   MORPHOLOGIES: *PLECTOSPIRA*

Plectospira myriandra (isolate 84–209 (ATCC64139))

A. Zoospores encysted in a mass at the tip of an evacuation tube exit from lobate sporangium.

B. Oogonium surrounded with antheridia and lobate sporangium.

C–D. Oogonium covered with several antheridia diclinously and monoclinously.

E. Regermination from encysted zoospore.

F. Oogonium germinated elongating germ tubes.

G. Parthenogenetically formed oospore. Sporangia can reach up to 25 μm wide; oogonia can be 15–28 μm in diameter; and encysted zoospores can reach 5.5–15 μm.

(Watanabe, 1987)

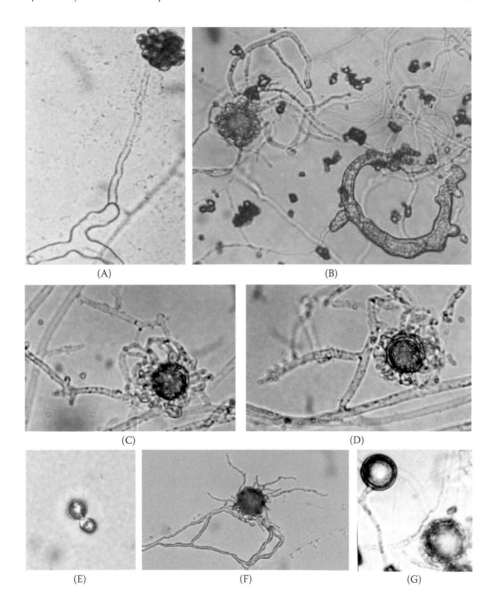

(A)

(B)

(C)

(D)

(E)

(F)

(G)

1.3 *PLECTOSPIRA MYRIANDRA*

A. Radial mycelial growth rates of *P. myriandra* (isolate 84–209 (ATCC64139)) after incubation for 24 hours on corn meal agar culture under different temperatures.

B. Zoospore discharge (number of encysted zoospores/12.6 mm^2) of *P. myriandra* under different temperatures.

C. Colony of *P. myriandra* (isolate 84–209, dried culture specimen).

(Watanabe, 1987)

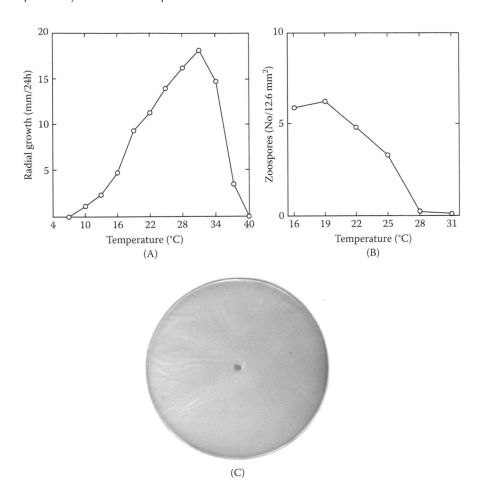

(A)

(B)

(C)

2 *Phytophthora*

2.1   DISEASES: OVERVIEW

Phytophthora diseases occur commonly in Japan, including pineapple heart rot (**A–B**) in Okinawa, *Phytophthora* diseases of orchid (**E**) in Chiba, and *Zanthoxylum piperitum* (**C–D**) in Nara.

Pineapple heart rot (**A–B**)

The disease started rotting internally, and the basal rotted leaves were readily removable. The disease was caused by both *P. nicotianae* var. *parasitica* and *P. cinnamomi*.

Phytophthora disease of orchid (**E**)

The disease was caused by *P. nicotianae* var. *parasitica*. The blighted lesions occur abundantly on leaves.

Phytophthora disease of Zanthoxylum piperitum (**C–D**)

The diseased plants were discolored and weakened. The root systems were poor and also darkened. The disease was caused by an unidentified *Phytophthora* species.

Stem blight of rose

The hydroponic culture of rose seedlings was damaged for the first time in 1968 in Chiba, and plants showed wilting, yellowing, and defoliation (**F**). In pathogenicity tests, the three-month-old rooted cuttings inoculated by burying the inoculum of *P. megasperma* in the soil near the healthy cuttings grown in pots or dipping the roots in the inoculum suspension were seriously damaged, especially when artificially wounded prior to inoculation by cutting a part of the roots. The new shoots and lower leaves of the inoculated plants initially wilted (**B**) and the plants finally died (Nagai et al., 1978). The fungus was isolated for the first time by the crown-rot pathogen of hollyhocks by Drechsler (1931).

2.2   DISEASES: PINEAPPLE, CATTLEYA, ROSE

A–B. Pineapple heart rots (**A**), and their basal rotted leaves removed (**B**).

C–D. Diseased, discolored *Zanthoxylum* plants with the poor root system (**D**).

E. Lesions of *Cattleya* around the inocula of *P. nicotianae* var. *parasitica*.

F. Defoliated lower leaves of naturally infected rose plants (Nagai et al., 1978).

G. Wilt in a young shoot and leaves after artificial inoculation under greenhouse conditions.

(A)   (B)   (C)

(D)   (E)

(F)   (G)

2.3   *PHYTOPHTHORA* DISEASES: PUMPKIN, FIG

A. Pumpkin seedling damaged, showing water-soaked symptoms on stems caused by *Phytophthora* sp.

B. Basal rot of young pumpkin at Wakayama, Japan.

C–D. Fig blight caused by *Phytophthora* sp. at Wakayama.

(A)

(B)

(C)

(D)

2.4 MORPHOLOGIES

Phytophthora megasperma: Sporangia were nonpapillated, elliptical, or ovate, and often proliferated internally and externally. Zoospores were differentiated inside sporangia and liberated through the open apical part. Encysted zoospores were spherical. Sexual organs are composed of oogonia and predominantly single paragynous (approximately 88%) antheridia. Oogonia were smooth, mostly globose. Occasionally observed oogonium stalks were funnel-shaped. Oospores are nearly plerotic. Antheridia were subglobose, terminal, or intercalary on the antheridial stalks. The fungus grew within the range 10–35°C on PDA, and its optimum temperature was at 25°C. No growth occurred below 5°C or above 38°C even after incubation for 20 days.

(*P. megasperma*: sporangia 23–70 × 18–42 µm. Zoospores encysted, 8.7–12.5 µm in diameter; oogonia 35–55 µm in diameter; oogonium stalks 7.5–120 × 10–25 µm; oospores 27.5–49 µm in diameter; antheridia 7.5–25 µm long, 10–20 µm in diameter)

P. nicotianae var. *parasitica:* Sporangia were ellipsoidal, well papillated, and nonproliferated. Sexual organs were composed of oogonia and single globose or subglobose antheridia, which are paragynous to oogonia. Oospores were aplerotic and chlamydospores were globose.

A. Growth (colony diameter in mm) of two *P. megasperma* isolates, P-A and P-B on PDA after incubation for three days at 5–35°C: they grew within the range of 10–35°C. The optimum temperature for mycelial growth was 25°C for both isolates. The fungus did not grow at 5°C or 38°C, even after incubation for twenty days.

(*P. nicotianae* var. *parasitica* sporangia 30–65 × 30–45 µm; oogonia 17.5–32.5 µm in diameter; oospores 14–30 µm in diameter; chlamydospores 22.5–40 µm in diameter)

(Drechsler, 1931; Nagai et al., 1978)

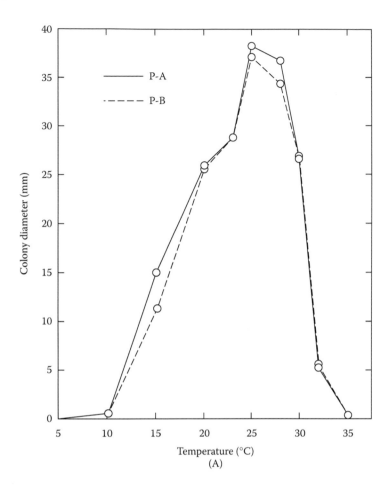

(A)

2.5   MORPHOLOGIES: 1. *P. NICOTIANAE* VAR. *PARASITICA*, AND 2. *P. MEGASPERMA*

1. *P. nicotianae* var. *parasitica* of *Zanthoxylum piperitum* isolate (**A**) and *Cattleya* isolate (**B–D**)

    A. Four papillate zoosporangia.

    B. Two papillate zoosporangia.

    C. Zoospore discharge.

    D. Globose chlamydospores.

2. *P. megasperma* of rose isolate (**E–J**)

    E–F. Nonpapillate sporangia without (**E**) and with inflation at base (**F**).

    G–H. External (**G**) and internal proliferated sporangia (**H**).

    I–J. Paragynous (**I**) and amphigynous antheridia (**J**).

(Nagai et al., 1978; Watanabe, 2010)

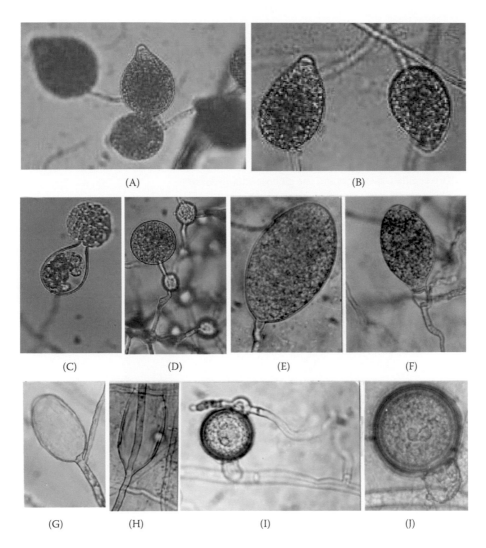

(A) (B)

(C) (D) (E) (F)

(G) (H) (I) (J)

2.6 *PHYTOPHTHORA* MORPHOLOGIES

Additional figures of *Phytophthora* on *Cattleya* (**A–D**) and *Phellodendron* isolate (**E**):

A. PDA colony of *Cattleya* isolates (91–8 and 91–16).

B. One papillate sporangium and three chlamydospores.

C. Zoospore discharge.

D. Chlamydospores.

E. Two papillate sporangia and one globose chlamydospore.

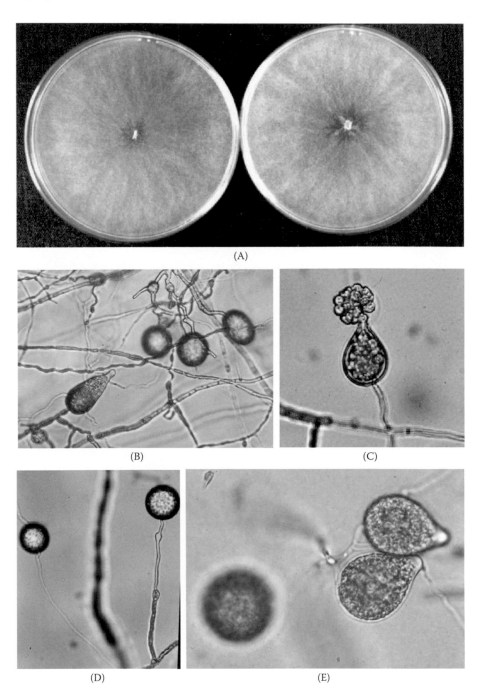

(A)

(B)

(C)

(D)

(E)

3 *Pythium*

3.1 DISEASES: OVERVIEW

Most seedling collapses before and after emergence may be due to pathogenic *Pythium* species. Young seedlings of various vegetables are often infected, damped-off, or stunted. Cucumber, tomato, and spinach are commonly diseased in Japan by *P. aphanidermatum, P. myriotylum,* and *P. ultimum.* Strawberry plants were often stunted in Japan and among various fungi associated with the roots. *P. ultimum* was the most pathogenic among them. At Okinawa, high-temperature favorite *P. deliense* is associated with bean seedling damping-off. Melon seedlings in central Japan have been often attacked by *P. splendens.* Soybean stems were also blighted basally and the roots were discolored brown. Brown-blotted root rot of carrot also occurred in Chiba and was caused by *P. sulcatum.* Summer-maturing carrots were diseased in the tap roots with the enlarged symptoms whereas winter-maturing carrots were diseased with brown-lenticel-like spots. Petiole rot of arrowhead (*Sagittaria trifolia*) and dasheen plants were caused by *P. myriotylum.*

P. carolinianum and *P. periplocum* were two *Pythium* species associated with flowering cherry seeds in nature, which is interesting to note. Various mature plants, commonly associated with *Pythium* spp., were often shriveled and their roots discolored.

In Paraguay, *Araucaria* saplings were often stunted with the poor root systems and associated with *P. splendens.* Black pepper seedlings in the Dominican Republic were also similarly diseased with *P. splendens.*

(Watanabe, 1981, 1983; Matsuda et al., 1998; Nagai et al., 1986, 1988)

3.2 DISEASES: DAMAGED CUCUMBER AND SOYBEAN PLANTS

A. Cucumber plants collapsed with *P. aphanidermatum* in the fields in Okinawa.

B. Cucumber seedlings inoculated with *P. aphanidermatum* (left, isolate 79–74; center, isolate 79–79) in relation to uninoculated healthy seedlings (right).

C. Soybean basal stem blighted by *P. aphanidermatum* in Tokyo.

D. Collapsed (left) and incompletely emerged kidney bean seedlings grown in the potted soil artificially infested with *P. deliense.*

(Watanabe, 1981, 1983)

(A)

(B)

(C)

(D)

3.3   DISEASES: DAMAGED CUCUMBER, TOMATO, AND SOYBEAN PLANTS

A–B. Cucumber (**A**) and tomato seedlings (**B**), 14 days old inoculated with, from left to right: *P. aphanidermatum* (isolate 74–901 (ATCC36443)), *P. myriotylum* (isolate 74-864-1 (ATCC36440)), *P. ultimum* (isolate 73–54), and uninoculated as control.

C. Diseased soybean roots inoculated with *P. myriotylum* (right), and uninoculated healthy roots (left, 24 days after inoculation).

P. aphanidermatum and *P. ultimum* were strongly pathogenic to cucumbers and tomatoes causing pre-emergence damping-off and severe stunting, but *P. myriotylum* was weakly or not pathogenic to these plants. However, inoculated soybean roots were severely lesioned.

(Watanabe, 1984)

(A)

(B)

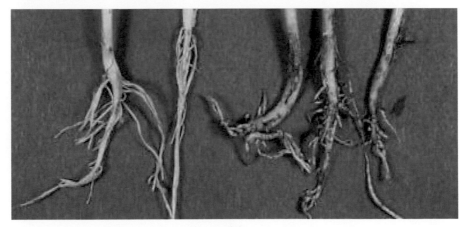

(C)

3.4   DISEASES: DISEASED *ARAUCARIA* AND BLACK PEPPER

A. Diseased *Araucaria* saplings with discolored leaves in the naturally infested field with *P. splendens* in Paraguay.

B. Damaged *Araucaria* roots with poor root systems.

C. Black pepper seedlings transplanted into artificially infested soil with *P. splendens* (right two pots) and noninfested soil (left two pots) in the Dominican Republic.

D. Two stunted seedlings in infested soil (left) and one healthy seedling in noninfested soil (right).

(Matsuda et al., 1998)

(A)

(B)

(C)

(D)

3.5 DISEASES: 1. DAMAGED MUSKMELON, AND 2. BROWN-BLOTTED ROOT ROT OF CARROTS

Muskmelons damaged by *Pythium splendens* in greenhouse soil and potted soil

A. Inoculation of the fungi to muskmelon plants in the field under greenhouse conditions.

B. Two collapsed plants inoculated with *P. splendens* (right) and one healthy plant (left).

C. Healthy seedlings (left) and collapsed inoculated seedlings with isolate 82–18 (right).

D. Healthy white roots (left), and diseased brown roots (right).

Brown-blotted root rot of carrots caused by *Pythium sulcatum*

E–F. Diseased taproots of summer-maturing carrots with enlarged lesions (**E**) and winter-maturing carrots with brown-lenticel-like spots (**F**).

(Watanabe, 1983; Uematsu et al., 1985; Nagai et al., 1986)

(A)

(B)

(C)

(D)

(E)

(F)

3.6   DISEASES: SPINACH, CHINESE CABBAGE, SOYBEAN, AND TOMATO SEEDLINGS

A–B. Spinach and Chinese cabbage seedlings damaged by *P. aphanidermatum* in the fields in Yamagata.

C–D. Basal stem rot of soybean and tomato caused by *P. aphanidermatum* in Tsukuba and Okinawa, respectively.

(Watanabe, 1983)

(A)   (B)

(C)   (D)

3.7 DISEASES: SOYBEAN SEEDLINGS AND PETIOLE ROT OF ARROWHEAD

A. Soybean seedlings in noninfested potted soil (as control, leftmost) and artificially infested with *P. aphanidermatum* (isolate 73–54 (ATCC36431)), *P. myriotylum* (isolate 74-864-1 (ATCC36440)), and *P. ultimum* (isolate 74–901 (ATCC36443)) from rightmost to left, 14 days after seeding.

B. Soybean seedling roots in noninfested soil (left two) and diseased roots in infested soil with *P. myriotylum* (right three roots).

C–D. Petiole rot of arrowhead (*Sagittaria trifolia*) caused by *P. myriotylum*. Note cracked and curled lesions on the petioles.

(Watanabe, 1983; Zenbayashi et al., 1985)

(A)

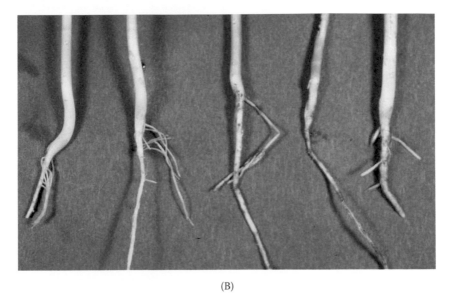

(B)

(C)   (D)

3.8   DISEASES: TOMATO AND SPINACH SEEDLINGS

A. Healthy and collapsed tomato seedlings grown in noninfested (as control, left), and infested soil with *P. ultimum* (isolate 74–901), 14 days after seeding.

B. Two plants each with healthy (left) and diseased tomato seedlings (right).

C. Healthy (right two plants) and diseased spinach seedlings (left three plants) grown in noninfested soil, and soil infested with *P. ultimum* (isolate 74–901), respectively.

(A)

(B)

(C)

3.9   MORPHOLOGIES: *P. APHANIDERMATUM*

The fungus favors high temperatures over 30°C, and distributes widely in Japan with air temperatures over 10°C. Sexual organs are observable within a few days, including immature oogonia with intercalary-formed antheridia on hyphae. Mature oogonia with aplerotic oospores and intercalary-formed antheridia were later observed.

(*P. aphanidermatum*: Sporangia 107–200 × 7–13.4 µm; vesicle 30–50 µm in diameter. Oospores encysted 10–12 µm in diameter. Oogonia 25–32.5 µm in diameter; oospores, 17.5–25 µm in diameter. Antheridia 10–22.5 × 10–12.5 µm).

(Watanabe, 1983, 1984)

A. *Pythium aphanidermatum* (isolate 73–54 (ATCC36431)).

Bottom row, left to right: cultures grown at 15°C and 25°C for 24 hours.
Top row, left to right: cultures grown at 28°C, 30°C, and 37°C for 24 hours.

B. *P. aphanidermatum* (left, isolate 73–54) and *P. deliense* (right, isolate 72–X108, ATCC38893).

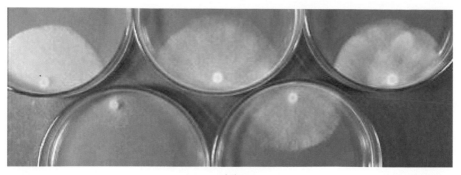

(A)

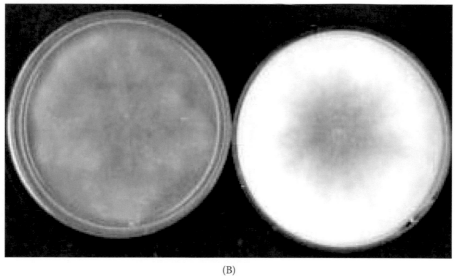

(B)

3.10   MORPHOLOGIES: *P. APHANIDERMATUM* (CONTINUED)

Detection of *Pythium aphanidermatum* from soil

A–B. Lobate sporangia and vesicles directly detected on water agar from soil in the plates (magnification, ×30).

C. Three vesicles with mature zoospores.

D. Zoospores discharged.

E. Two immature oogonia with intercalary antheridia formed on hyphae.

F. Mature oogonium bearing an aplerotic oospore, and typically intercalary globose antheridium.

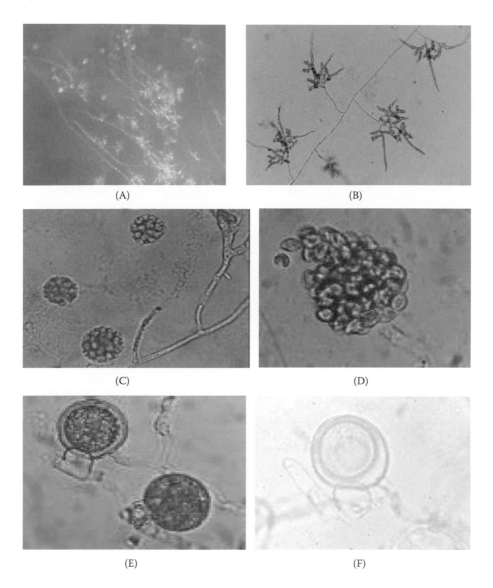

(A)

(B)

(C)

(D)

(E)

(F)

3.11   MORPHOLOGIES: *P. DELIENSE*

Pythium deliense compared with *P. aphanidermatum*

A–B. Oogoniumphores of *P. deliense* characteristically bending toward antheridia.

C–D. *P. deliense* (**C**) is smaller than *P. aphanidermatum* (**D**) in oogonium size.

P. aphanidermatum morphologically differentiated from *P. deliense* in which oogoni-umphores are bent toward antheridia; oospore sizes are smaller and favor rather lower temperatures.

(*P. deliense*: sporangia 30–200 × 7.2–7.5 µm, vesicle 30–50 µm in diameter. Oogonia 15–26.3 µm in diameter; oospores, 13.3–20.7 µm in diameter. Antheridia 5–22.5 × 2.2–8.6 µm)

(Watanabe, 1981)

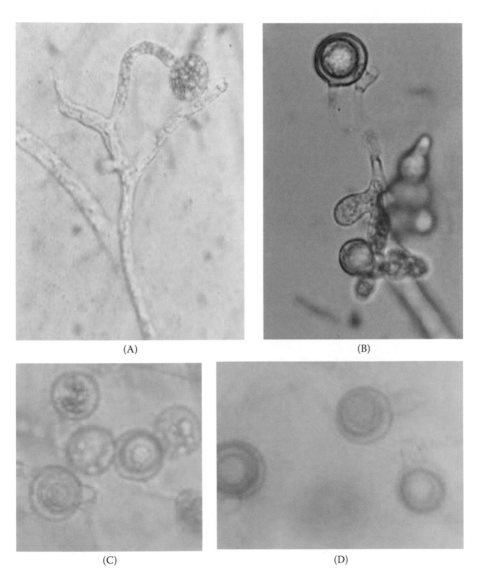

(A)

(B)

(C)

(D)

3.12   MORPHOLOGIES: *P. INTERMEDIUM*

A. A small sporangium basipetally formed, just below the large terminal sporangium.

B–C. Four deciduous catenulate sporangia (**B**), two, and eight catenulate sporangia (**C**).

D. A total of 14 catenulate sporangia and two detached single sporangia.

E. Sporangium germinated elongating one germ tube.

F. PDA colony of isolate 90–116.

G. Radial mycelial growth rates of four isolates of *P. intermedium* after incubation for 24 hours on PDA cultures under different temperatures.

Globose sporangia are formed basipetally in chains (up to 23) that are deciduous: no sexual stage reported. They are isolated from various plant roots in Japan, distributed in the mountainous soils and weakly pathogenic to cucumber seedlings. Encysted zoospores are rarely observed.

Note: zoospores are discharged from 10% to 20% of sporangia (Stanghellini et al., 1988).

(Watanabe, 1983; Watanabe et al., 1998)

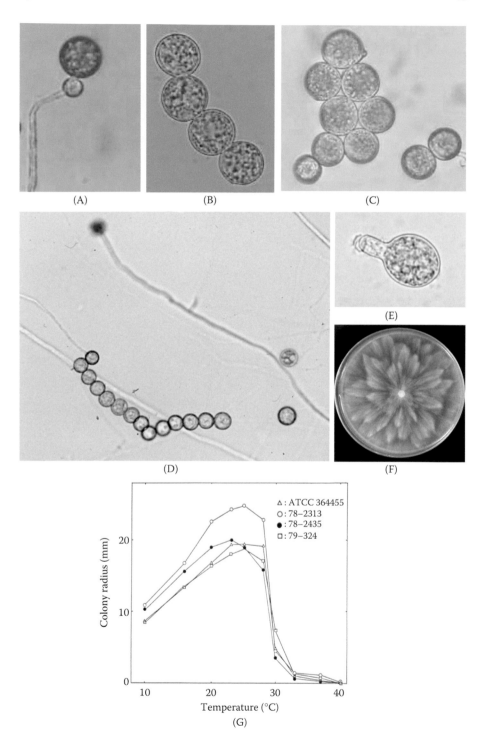

(A) (B) (C)

(D) (E) (F)

(G)

3.12.1 MORPHOLOGIES AND DISEASES: SOYBEAN, TOMATO, AND DASHEEN PLANTS

A. Healthy soybean and tomato seedlings (left two pots) grown in noninfested soil versus diseased seedlings in the infested soil with *P. myriotylum* (right two pots).

B. Diseased dasheen plant infected with *P. myriotylum*.

C. Tuber with water-soaked discolored roots.

(Watanabe, 1981; Nagai et al., 1988)

(A)

(B)   (C)

3.12.2   MORPHOLOGIES AND DISEASES: *P. MYRIOTYLUM,* DASHEEN PLANT

The fungus is characterized by lobate sporangia, which form long exit tubes and terminal vesicles, and oogonia bearing aplerotic oospores surrounded by diclinous crook-necked antheridia. It favors high temperatures over 30°C for growth.

Root diseases of strawberry, petiole rot of arrowhead, and various diseases occur in Japan.

(*Pythium myriotylum*: Sporangium exit tubes up to 450 µm long; vesicles 40–55 µm in diameter. Oogonia 28.7–37.5 µm in diameter; oospores 20–32.5 µm in diameter; Antheridia ca. 15 µm long, 7.5 µm wide)

(Watanabe, 1977; Zenbayashi et al., 1985; Nagai et al., 1988)

A. *P. myriotylum*: mycelial growth rates of the dasheen isolate (isolate 75–26) on potato sucrose agar media for 24-hour incubation at various temperatures.

(Nagai et al., 1988)

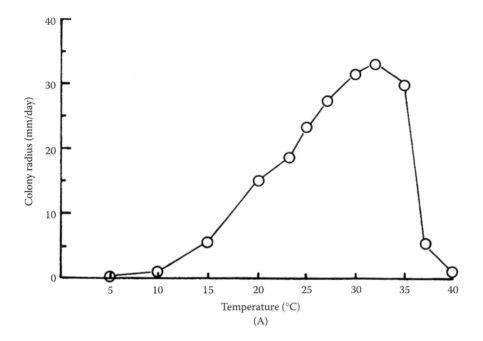

(A)

3.13 MORPHOLOGIES AND DISEASES: *P. MYRIOTYLUM*

A–C. Oogonium including undifferentiated oosphere with a few antheridia from diclinous antheridiumphores (**A–B**) and differentiated oosphere and lobate sporangium (**C**).

D. Oogonium bearing an aplerotic oospore covered with antheridia.

E. Vesicle formed at the end of an elongated exit tube from lobate sporangium.

F. Zoospore discharge from vesicle.

G. Appressoria.

H. PDA colony of isolate 83–193.

(Watanabe, 1977; Zenbayashi et al., 1985; Nagai et al., 1988)

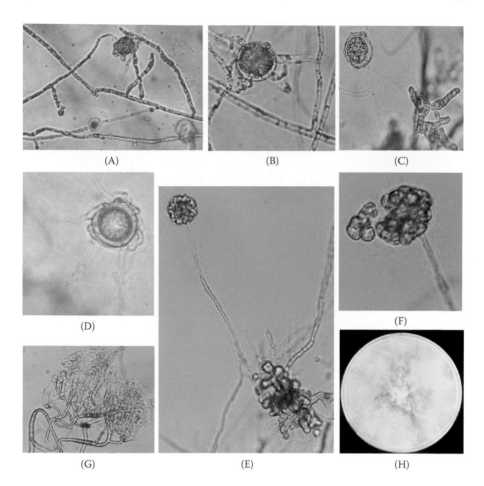

(A) (B) (C)

(D)

(F)

(G) (E) (H)

3.14 MORPHOLOGIES: *P. SPLENDENS*

A–B. Hyphae bear two (**A**) and four terminal sporangia (**B**).

C. PDA colony of isolate 85–18.

D. Four *P. splendens* isolates tested grew best at 28°C on PDA. The colony radius reached 35–38 mm for 24-hour incubation.

(The sporangia were over 40 μm in diameter. Sexual organs were not formed.)

(Watanabe, 1981)

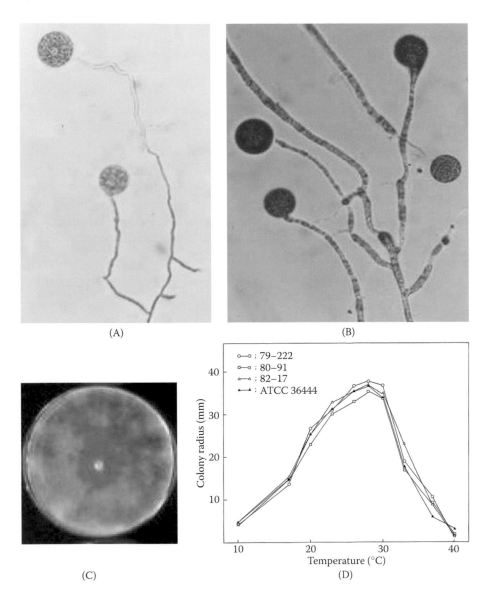

(A)

(B)

(C)

(D)

3.15 MORPHOLOGIES: *P. SULCATUM*

A. The hyphal swellings are catenulate, globose, and often peanut-shaped. No zoospores are formed.

B–D. Oogonia bearing plerotic oospores surrounded by monoclinus or diclinus, crook-necked, furrowed antheridia.

E. Oospore is aplerotic with a central reserve oil globule.

F. Twelve-day-old PDA colony in a 9-cm plastic Petri dish at 28°C.

G. Radial mycelial growth rates of *P. sulcatum* (isolates A–1 = 83–161, ATCC20997; and A–4 = 83-160, ATCC200998) after incubation for 24 hours on PDA under different temperatures.

(*Pythium sulcatum*: conidia 20–30 μm in diameter. Oogonia 15–28.8 μm in diameter; oospores 13.7–22.5 μm in diameter. Appressoria 2.2–6.3 μm in diameter)

(Nagai et al., 1986; Pratt and Michell, 1973; Watanabe et al., 1986)

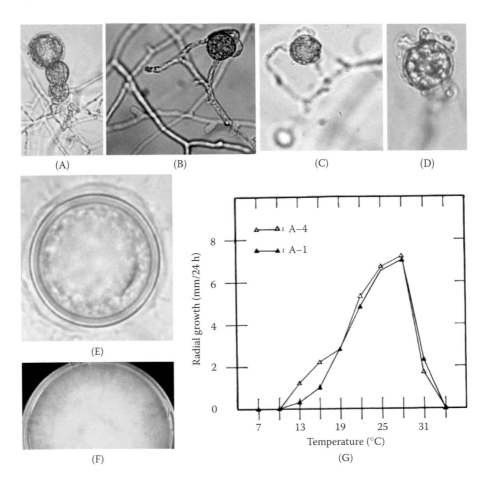

(A)　(B)　(C)　(D)

(E)

(F)

(G)

3.16   DISEASE AND MORPHOLOGIES: STRAWBERRY STUNT DISEASE AND *P. ULTIMUM*

The strawberry plants were grown in vinyl houses or sometimes vinyl tunnels, for cold protection together with polyethylene film mulch in the 1970s.

A disease of strawberry plants known as "stunt," characterized by poor growth, root deterioration, and low yields, was serious in Japan. Similar diseases occurred worldwide, including black root complex in California (Wilhelm et al., 1972), black root in Michigan (Strong and Strong, 1931), root rot in Illinois and Canada (Hildebrand, 1934; Nemec, 1970; Truscott, 1934), and progressive decline in Italy (D'ercole and Canova, 1974). Various fungi were implicated as the pathogens of these diseases.

Of 58 fungus genera identified among 2011 fungus isolates from "stunted" and healthy strawberry plants from May 1973 to December 1974 in Japan, 60% were species of *Fusarium, Pythium,* and *Rhizoctonia.*

Among 291 isolates of 14 *Pythium* species identified, *P. sylvaticum* complex, *P. ultimum, P. spinosum,* and *P. oedochilum* were the dominant species. *P. ultimum* was a primary pathogen of the strawberry stunt disease among several species tested at temperatures below 20°C. In strawberry plants grown in noninfested control soil, an average of 35.5% of 246 roots checked were discolored; whereas 63.4% of 250 roots were discolored in soil infested with *P. ultimum* 50 days after being planted in a greenhouse at 20°C–38°C. Furthermore, fresh total plant weights in noninfested control soil were 24 g, but 10 g in the *P. ultimum*-infested soil 60 days after being planted at 20°C (Watanabe et al., 1977).

A. Strawberry stunt disease caused by *P. ultimum.* Healthy (left) and diseased plants (right) that occurred in Shizuoka.

B. *P. ultimum* (isolate 83–412) from forest soil in Mt. Fuji.

C. *P. ultimum* (isolate 74–901 (ATCC36443)) on PDA cultures.

Bottom row, left to right: cultures grown at 15°C and 25°C for 24 hours.

Top row, left to right: cultures grown at 28°C, 30°C, and 37°C for 24 hours.

(Watanabe et al., 1977)

(A)

(B)

(C)

3.17 *P. CAROLINIANUM* AND *P. PERIPLOCUM* ASSOCIATED WITH FLOWERING CHERRY SEEDS

Two *Pythium* species were found as flowering cherry seed associates, during the study on seed-borne pathogenic fungi of forest trees.

Five hundred seeds each of two cherry species, *Prunus lannesian* var. *speciosa* (Ōshimazakura in Japanese) and *P. jamasakura* (Yamazakura) were assayed within one year after harvest in Asakawa, near central Tokyo, where about 1800 flowering cherry trees of 215 species and varieties for collection, protection, and the revivification have been planted.

Seed-associated *Pythium* may be noteworthy, but has not been previously reported in Japan and other countries, and may be related to the disappearance of young cherry seedlings under natural renewal. These fungi were also detected from forest soils together with seven other *Pythium* species. Sporangia of *P. carolinianum* are papillate, germinated directly or indirectly, and often proliferated internally. No sexual organs were formed.

In *P. periplocum* lobate sporangia, echinulate oogonia and aplerotic oospores were formed. Both *P. carolinianum* (isolate 85–54 (ATCC66260)) and *P. periplocum* (isolate 85–53 (ATCC66262)) grew best at 28°C (Section 3.19, Figure C).

In inoculation to potted seedlings, *P. carolinianum* showed pathogenicity (Figure A) (Watanabe, 1988; Watanabe et al., 1987).

A. Healthy and collapsed flowering cherry seedlings grown in the noninfested (left) and infested soil with *Pythium carolinianum* (right), respectively.

B. Flowering cherry seeds (six seeds each of var. Ōshimazakura and Yamazakura) collected at Experiment Forest, FFPRI, Tama.

(A)   (B)

3.18   *P. CAROLINIANUM* AND *P. PERIPLOCUM*: INOCULATED JAPANESE BLACK PINE SEEDLINGS

A. *P. carolinianum* (left, isolate 85–54 (ATCC66260)) and *P. periplocum* (right, isolate 85–53 (ATCC66264)) in one-day-old PDA cultures.

B. Healthy and collapsed flowering cherry seedlings grown in the noninfested (left) and soil infested with *Pythium carolinianum* (center, isolate 85–54), and *P. periprocum* (right, isolate 85–53), respectively.

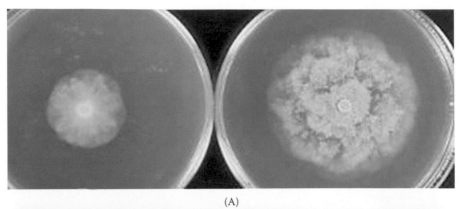

(A)

(B)

3.19   *P. CAROLINIANUM* AND *P. PERIPLOCUM* ASSOCIATED WITH FLOWERING CHERRY SEEDS: PDA CULTURES

A–B. *P. carolinianum* (isolate 85–54) (**A**) and *P. periplocum* (isolate 85–53) in nine-day-old PDA cultures (**B**).

C. Radial mycelial growth rates of *P. carolinianum* and *P. periplocum* after incubation for 24 hours on PDA at different temperatures.

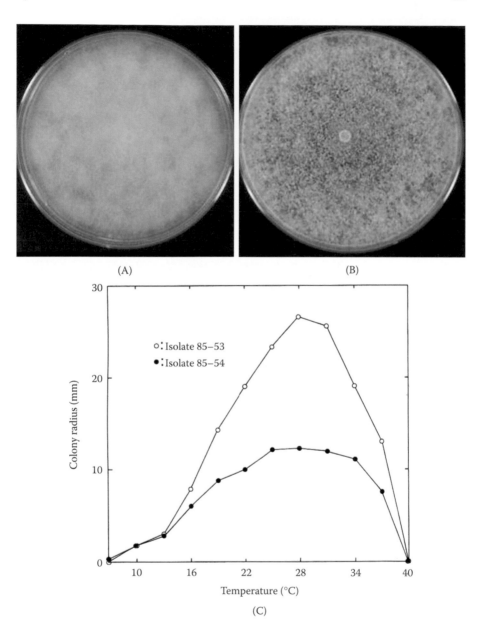

(A)

(B)

(C)

3.20   *P. CAROLINIANUM* AND *P. PERIPLOCUM* ASSOCIATED WITH FLOWERING CHERRY SEEDS: MORPHOLOGIES

Morphologies of *P. carolinianum* (**A–E**) and *P. periplocum* (**F–H**)

A–B. Apiculate (**A**) and nonapiculate (**B**) sporangia.

C. Zoospore discharge.

D–E. Internally (**D**) and externally (**E**) proliferated sporangia.

F–H. Echinulate oogonia with aplerotic oospores, and lobate sporangia (**G–H**).

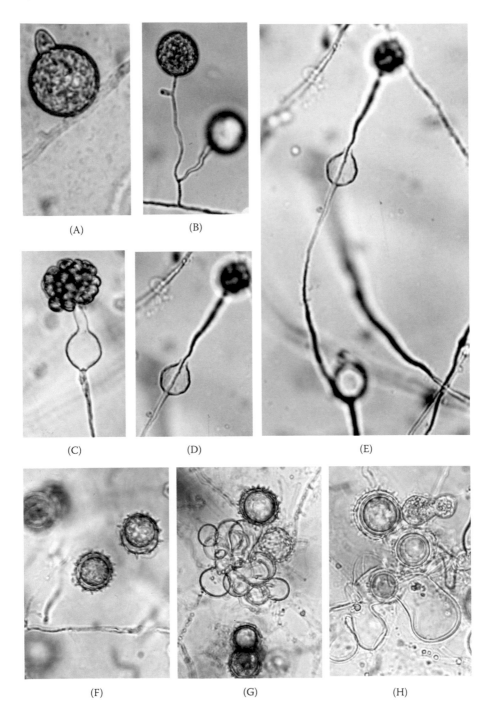

(A) (B)

(C) (D) (E)

(F) (G) (H)

3.21 PATHOGENICITY: *P. SPLENDENS* AND FOUR MELON ROOT-ASSOCIATED FUNGI

The pathogenicity of four *P. splendens* isolates was demonstrated in inoculation tests to potted muskmelon (4 cvs). Collapsed seedlings, and healthy and diseased roots were shown at 55 days after inoculation. Fungi were inoculated by mixing disintegrated mycelia harvested from potato dextrose broth culture with potted soil (one plate culture/500 g potted soil) (Section 3.5, Figures A to D, p. 31; Watanabe, 1983).

TABLE 3.1
Pathogenicity of *Pythium splendens* (isolates 82–18, 82–28, 82–34, and ATCC 36414) to Potted Muskmelon Seedlings (4 cvs)

Treatment	Amus	Sun-Rise	Super-Melon	Green Pearl	Total
82–18	9/12	3/12	5/9	3/10	20/45
82–28	5/10	1/9	7/10	1/9	14/38
82–34	5/9	4/10	6/10	9/10	24/29
ATCC36414	1/10	1/10	1/10	1/10	4/40
Control	0/10	0/10	0/10	0/10	0/40

Comments: Pre-damping-off, and collapse after emergence occurred in some seedlings. Isolate 82–34 was the most virulent among four isolates tested.
Source: Watanabe, 1983; Onogi et al., 1984

TABLE 3.2
Pathogenicity of Melon-Associated Fungi: *Monosporascus cannonballus, Pythium splendens* and *P. sylvaticum,* and *Rhizoctonia solani* to Melon Seedlings (3 cvs)

	No. of Diseased Seedlings/No. of Emerged Seedlings			
Fungi Tested	Amus	Arlus	Super	Total
Monosporascus cannonballus (isolate 79–51)	4/10	1/12	10/12	15/34
Pythium splendens (isolate 82–46)	11/12	7/12	11/12	29/36
P. sylvaticum (isolate 82–16)	11/11	3/12	11/12	25/35
Rhizoctonia solani (isolate 82–96)	2/12	3/12	12/12	17/36
Control	4/12	0/12	5/12	9/36

The inoculated seedlings collapsed or damaged with any necrotic symptoms were judged as diseased plants without definite etiological studies.
Source: Watanabe et al., 1983

3.22 MORPHOLOGIES: APPRESSORIA OF *PYTHIUM* SPECIES

A. *Pythium aphanidermatum*　　**B.** *P. glaminicola*　　**C.** *P. conidiophorum*
D. *P. dissimile*　　**E–F.** *P. myriotylum*　　**G.** *P. sylvaticum*
H. *P. sulcatum*

Appressoria of *Pythium* species vary in shape and characteristics in the respective species.

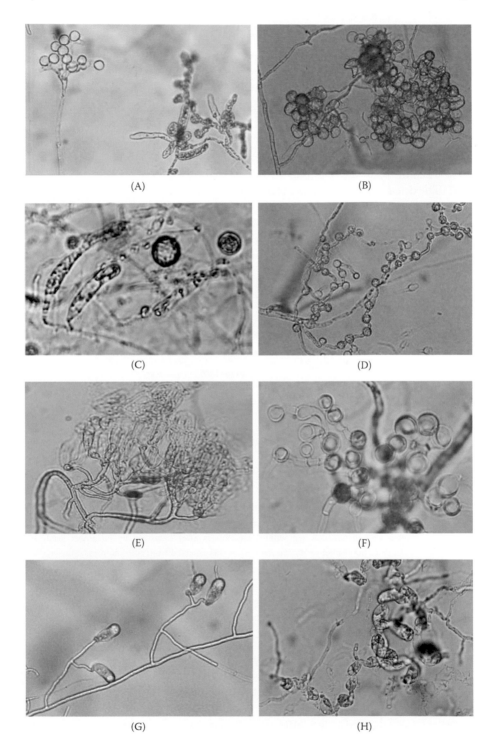

(A)

(B)

(C)

(D)

(E)

(F)

(G)

(H)

3.23   *P. APHANIDERMATUM* DETECTED FROM SOIL

Pythium species, including *P. aphanidermatum, P. myriotylum, P. splendens,* and *P. ultimum,* have been found commonly on plants and soils worldwide. *P. aphanidermatum* was detected from Japanese soil in the following assay procedures:

1. Soil sample (10 g) are placed in a 9-cm Petri dish, subsequently distributing cucumber seeds (10 seeds/plate).
2. Watered (6 mL/plate) and mixed.
3. The plate is incubated at 30°C over 12 hours.
4. Then the seeds are removed, washed, and air-dried.
5. The dried seeds (2/plate) are plated on 2% water agar.
6. Petri's solutions poured over them (4 mL/plate).
7. The plate is incubated at 30°C–35°C for over 12 hours.
8. Morphologies of sporangia, vesicles, and zoospores may be observed within 24 hours, and sexual organs and oospores within 48-hour incubation (Section 3.10, p. 41).

TABLE 3.3
Detection Frequency of *Pythium aphanidermatum* in Tsukuba Soils, Japan at 30°C or 35°C by a Simple Assay Method

	Detection Frequency[a]			
	Sample A[b]		Sample B	
Incubation (hr)	30°C	35°C	30°C	35°C
	1	2	1	2
1	0/20	0/20	0/20	0/20
3	0/20	0/20	0/20	1/20
6	0/20	0/20	5/20	4/20
12	5/20	2/20	7/20	7/20

[a] Number of seeds yielding *P. aphanidermatum*/number of seeds tested.

[b] Samples A and B were collected in January and August 1983, respectively.

Note: *P. aphanidermatum* was detected within three hours after treatment by cucumber seeds in Sample B. In 12 hours after treatment, the fungus was detected from both samples.

Source: Watanabe, 1984

3.24   TEMPERATURE RESPONSES OF *P. APHANIDERMATUM* AND *P. DELIENSE*

A. A total of 15 Japanese isolates of *P. aphanidermatum* were tested, including five isolates from the Ryukyu Islands (solid circle), ten (blank circle) from the Tohoku district, and six *P. deliense* isolates from the Ryukyu Islands (solid triangle), showed best growth at temperatures above 30°C, with the optimum at around 35°C on PDA in radial mycelial growth rates (mm) after incubation for 24 hours.

Both species were collected from fields of strawberries and sugar canes.

P. aphanidermatum and *P. deliense* grew well at temperatures above 30°C with the optimum temperature at around 35°C.

(Watanabe, 1981)

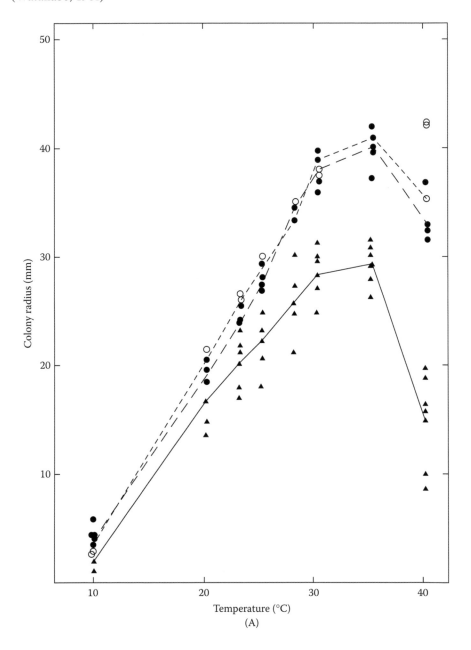

(A)

3.25   TEMPERATURE RESPONSES OF
SWISS ALPS *PYTHIUM* ISOLATES

A. Some ten species were detected from the Swiss Alps soils (Watanabe et al., 1998), and they showed different temperature responses. Most species favored 25°C, but three species favored temperatures below 25°C. Species that favored high temperatures were not detected in this study. Among 49 isolates of various *Pythium* species in the Swiss Alps assayed, one isolate grew best at 31°C, nine grew best at 28°C, five (e.g., isolate 90–109) at 22°C, but most of the others grew best at 25°C.

Among these cultures, the following representative species are deposited at ATCC with their respective ATCC accession numbers:

Isolate	Species	Accession Number
90–104	*P. indigoferae*	ATCC200701
90–113	*P. indigoferae*	ATCC200702
90–107	*P. intermedium*	ATCC200704
90–109	*P. intermedium*	ATCC200705
90–128	*P. intermedium*	ATCC200706
90–131	*P. rostratum*	ATCC200620
90–124	*P. rostratum*	ATCC200703
90–141	*P. sylvaticum*	ATCC200621
90–142	*P. sylvaticum*	ATCC200622
90–143	*P. sylvaticum*	ATCC200623
90–122	*P. torulosum*	ATCC200638
90–105	*P. torulosum*	ATCC200624

(Watanabe et al., 1998)

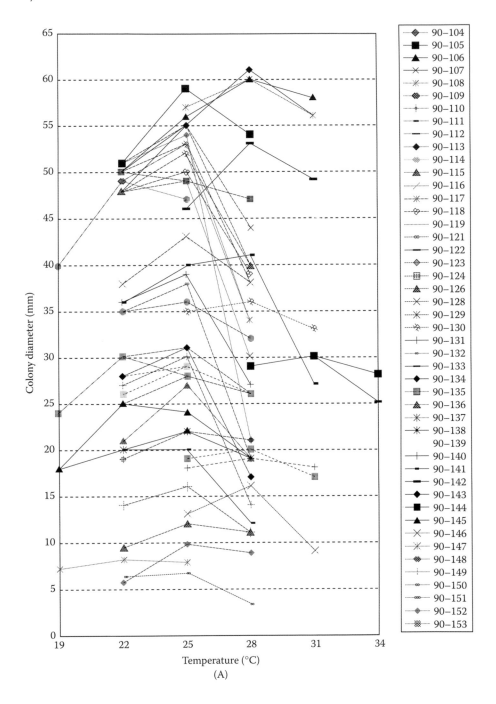

(A)

3.26   TEMPERATURE RESPONSES OF *PYTHIUM* SPECIES

TABLE 3.4

Temperature–Growth Relations of 30 *Pythium* Isolates from Strawberries of Japan and Sugarcane of Taiwan Based on the Radial Mycelial Growth Rates (/day)

	Mycelial Growth Rate (/day)		
Temp (°C)	Slowest (<20 mm)	Intermediate (20–30 mm)	Fastest (>30 mm)
25–28	*P. carolinianum* (5)*	*P. catenulatum* (99)	*P. sylvaticum* (866)
	P. debaryanum (114)	*P. intermedium* (483–2)	*P. ultimum* (42, 901)
31	*P. inflatum* (118)	*P. acanth.-P. oligand.* (42)	*P. echinulatum* (29, 822)
	P. paroecandrum (827)	*P. afertile* (663)	*P. spinosum* (16)
	P. torulosum (236)	*P. angustatum* (257)	*P. splendens* (192)
		P. gramin.-P. arrhen. (116)	
		P. inflatum (117)	
34–37	*P. inflatum* (51)	*P. gramin.-P. arrhen.* (115)	*P. aphanidermatum* (54)
			P. deliense (103, 130)
			P. myriotylum (864–1)
			P. oedochilum (165)

Note: *Isolate numbers:
    5 (73–5, ATCC36434)
    99 (772–X99, ATCC38892)
    866 (74–866, ATCC36442)
    483–2 (73-483-2, ATCC36445)
    42 (73–42, MAFF425485)
    901 (74–901, ATCC36443)
    29 (73–29, ATCC388911)
    822 (74–822, ATCC36437)
    827 (74–827, ATCC36432)
    663 (74–663, ATCC36439)
    16 (73–160, ATCC36438)
    192 (73–192, ATCC36444)
    117 (72–X117, ATCC38894)
    51 (73–51, ATCC36436)
    54 (73–54, ATCC36431)
    864–1 (74-864-1, ATCC36440)
    165 (74-864-1, ATCC36433)
Source: Watanabe, 1978 (modified)

Most *Pythium* species tested among isolates from strawberries in Japan and sugar cane in Taiwan favored high temperatures for growth at around 30°C. *P. aphanidermatum, P. deliense, P. myriotylum,* and *P. oedochilum* favored temperatures over 35°C (Watanabe, 1978, modified).

3.27 *PYTHIUM* SPECIES IN JAPAN: ISOLATION, TYPES OF SPECIES, NUMBER OF SPECIES PER DISTRICT, AND NUMBER OF PROPAGULES PER GRAM OF SOIL

Plants often did not emerge because of unknown reasons, or the seedlings were dwarfed, wilted, shriveled, or more or less damaged after emergence. According to the location, such phenomena occur very often, and thus, soil disinfection has been practiced before planting in culture practices.

Fresh potato cubes were buried in soil samples collected from 27 locations in the Tohoku district; then, their degradation in soil was checked. No cubes were disintegrated (their disintegration rate = 1) in eight locations just like the discs planted in the autoclaved soil as a control. However, the cubes' disintegration rates were severe (3.0–4.0) in five locations; moderate (2.0–3.0) in three locations; and weak (1.2–1.8) in eight locations. In Sample 17 from Aomori, cubes were completely disintegrated (the disintegration rate = 4) (Figure A, p. 75). Potato cube disintegration may be due to microbial activities; particularly, *Pythium* might be partially concerned in these activities.

Only *Pythium debaryanum* was reported to be associated with cucumber seedling, corn roots, and several other seedlings as early as 1906 in Japan (Shirai, 1906). Now some 100 species have already been reported worldwide (Plaats-Niterink, 1981).

Pythium species have been found everywhere often together with *Fusarium* and *Rhizoctonia* species in Japan. These fungi were associated with the disorders of various young seedlings, and were also readily detected or isolated from the nearby soils by baiting technique with cucumber seeds or other substrates. More than ten species were found commonly in the soils, including pathogenic species such as *P. aphanidermatum, P. irregulare, P. myriotyum, P. spinosum, P. splendens, P. sylvaticum, P. torulosum, P. ultimum,* and *P. vexans.*

A total of 222 soil samples, including cultivated soils planted with around sixty types of crops and uncultivated soils of various habitats (that is, forest soils) were collected from seven districts in Japan. Soil pH of these samples ranged from 4.4–8.3.

Pythium was isolated by cucumber seeds as a trapping substrate. A total of 2116 isolates, 207–415 isolates per district, were obtained and identified into at least 24 taxa, 11–16 per district including H-Zs (a provisional *Pythium* group with zoospore formation from hypha-like sporangia, but without sexual organs). *Pythium* was not isolated from 23 samples. These soils may be free from *Pythium* or contain too few in their populations to be trapped, but others yielded one to seven taxa per sample. Average numbers of taxa per sample ranged from 1.9 in Shikoku to 4.2 in Kinki, with a ground average of 2.8.

(Watanabe, 1985; 1988a, b)

3.28 DETECTION OF *P. APHANIDERMATUM* FROM VARIOUS AREAS IN JAPAN

Pythium spinosum and *P. sylvaticum* are two main species commonly found all over Japan. However, in the mountainous areas, *P. intermedium* and *P. torulosum* are also commonly found. Detection from the piedmont soils in Mt. Fuji and also from the Swiss Alps soils are such examples in my experience.

Pythium species were detected from all over Japan, and not detected in a few soil samples assayed for 100 areas in Japan. Most of the soil samples assayed contained some 1000 propagules per gram of soil. The more abundant propagules were found in the soils near the host plants or cultivated soils rather than uncultivated soils.

In the southern Japanese areas with high-temperature climates, *P. aphanidermatum, P. deliense,* and *P. myriotylum* are commonly present as compared with other areas. For example, *P. aphanidermatum* was detected in 52.4% of the Ryukyu Islands with the average temperatures above 20°C, 48% of the Kyushu district with 14–17°C, and in four out of 27 samples assayed in the Tohoku district with 10°C, and no fungus was detected in 21 samples from Hokkaido with below 8°C.

P. deliense appears to be limited to the Ryukyu Islands in the distribution, and was isolated from pumpkin, tomato, and bean roots in addition to soil samples.

P. myriotylum was isolated from strawberry, dasheen, and Chinese arrowhead in the Kanto district, but it was also found in Kitami, Hokkaido. In soil samples from the Ryukyu, these propagules were commonly present.

Populations of *Pythium* spp. were assayed in soil samples from Tohoku and the Ryukyu district using a modification of Waksman's method.

Out of 27 samples from the Tohoku district, nine samples each yielded more than 201 propagules per gram of soil; eight samples each yielded 51–200 propagules; and ten samples each yielded fewer than 50 propagules. Out of 21 soil samples from the Ryukyu Islands, three samples each yielded more than 201 propagules per gram of soil; nine samples each yielded 51–200 propagules; and nine samples each yielded fewer than 50 propagules.

Dominant species comprising these propagules in Tohoku soils were *P. sylvaticum* and *P. spinosum*, and in the Ryukyu, it was *P. aphanidermatum*.

The highest propagule number, an average of 1039 per g of soil, was recorded in Okinawa soil. All of the propagules belonged to *P. deliense*.

More than ten species were isolated from strawberry roots in Japan, and from sugar cane roots in Taiwan. *P. ultimum* was commonly found in strawberry environments and pathogenic, causing the stunt disease. *P. catenulatum* was unique in the relation to the poor ratooning of sugar cane in Taiwan (Watanabe, 1985).

3.29 DETECTION OF *PYTHIUM* FROM VARIOUS AREAS IN JAPAN

A. Degradation of potato cubes in soil. A total of 27 samples from the Tohoku district assayed (the minimum disintegration rate = 1 in control, the maximum = 4).

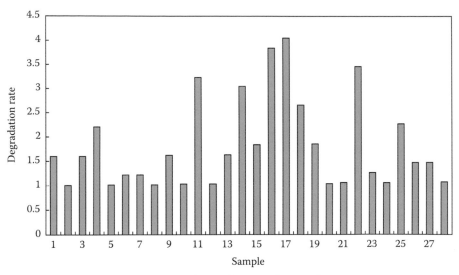

Degradation of potato cubes in soil

(A)

3.30   DETECTION AND ISOLATION OF *PYTHIUM*

Nine dominant *Pythium* taxa are listed with numbers of samples yielding the respective fungi in the seven assayed districts (Table 3.5).

TABLE 3.5
Dominant *Pythium* Taxa and Numbers of Soil Samples Yielding the Respective Fungi in the Seven Locations of Japan

Pythium species	Ryukyu	Kyushu	Shikoku	Kinki	Kanto	Tohoku	Hokkaido
P. aphanidermatum	12	13	4	10	1	6	0
P. dissotocum	1	0	2	2	9	2	0
P. irregular	1	4	3	13	7	4	3
P. spinosum	3	8	8	5	10	15	5
P. sylvaticum	13	22	28	14	32	23	16
P. torulosum	1	1	2	8	6	3	9
P. ultimum	4	17	10	12	5	6	6
P. vexans	11	1	4	5	3	1	2
H-Zs	3	2	1	18	8	2	4
Total samples	21	25	35	28	63	28	22

P. sylvaticum was predominant in most districts except Kinki where H-Zs was predominant. *P. aphanidermatum* was predominant in Ryukyu, Kyushu, and Kinki.

P. irregulare was dominant in Kinki, *P. spinosum* in Kanto and Tohoku, and *P. ultimum* in Kyushu, Shikoku, and Kinki.

Pythium was isolated in 16 out of 55 samples from uncultivated areas, but ten taxa in total and an average of 1.3 taxa per sample were found in these areas. *Pythium* was not isolated in seven out of 167 samples; 24 taxa in total and 3.2 taxa per sample were found. *P. sylvaticum* was always predominant in both cultivated and uncultivated areas, but following this species, *P. ultimum, P. spinosum,* and *P. aphanidermatum* were dominant in cultivated areas. In contrast, *P. intermedium, P. torulosum,* and *P. periplocum* were dominant in uncultivated areas.

In eggplant and lettuce fields, a total of 11 taxa (4.3–5 taxa per sample) were found, whereas only five taxa in total (1.7 per sample) were found in tomato field soils, *P. sylvaticum,* and *P. ultimum* were dominant.

At least 25 species were detected or isolated from soils in Japan. The number of detected species varied with districts ranging from ten species in Hokkaido to 17 species in Kyushu. *P. sylvaticum* was the most widely distributed, being detected in 125 out of 182 soil samples collected between 1978 and 1984. Other dominant species were *P. spinosum, P. aphanidermatum,* and *P. ultimum* in descending order.

DISTRIBUTION RATE

If the sample number of the fungus detected to the total number of samples assayed is defined as the distribution rate, the rate of *P. sylvaticum* was 68.7% and the subsequent three dominant species ranged from 24% to 27%. The dominant species in different soils are illustrated, together with the distribution rate (Figure A, p. 79).

The number of species present in a soil sample differed with samples. Of 182 samples assayed, an average of 2.6 species per sample was obtained. However, as many as six species were detected per sample in five samples. The average yield was 2–3.4 species per sample of each district.

3.31   DOMINANT *PYTHIUM* TAXA IN DISTRICTS OF JAPAN

A. *Pythium* species in districts of Japan with distribution rates (the sample number of the fungus detected to the total number of samples assayed) based on samples collected in 1978–1984.

(*Pythium* species abbreviated: Pa = *P. aphanidermatum*, Pdis = *P. dissotocum*, Pech = *P. echinulatum*, Pirr = *P. irregulare*, Ppul = *P. pulchrum*, Pros = *P. rostratum*, Pspi = *P. spinosum*, Pspl = *P. splendens*, Psyl = *P. sylvaticum*, Ptor = *P. torulosum*, Pu = *P. ultimum*, Pv = *P. vexans,* H-Zs = A group of fungi forming hypha-like sporangia but no sexual organs formed)

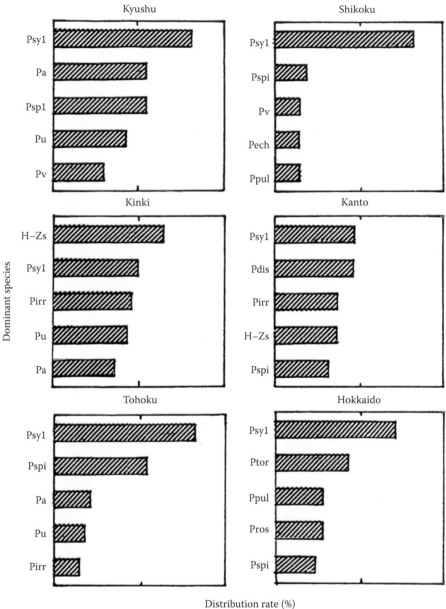

Distribution rate (%)

(A)

4 Zygomycetes

4.1   DISEASE: *RHIZOPUS* ROTS OF MULBERRY-GRAFTED SAPLINGS

Diseased mulberry grafted saplings caused by *R. oryzae* (**A–C**) and the morphology (**D–I**). Echinulate sporangium (**D**), subglobose columella (**E**), sporangiophore with basal rhizoid and collapsed apical sporangium (**F**), colony on potato-sucrose agar plate (**G**), chlamydospores (**H**), and sporangiospores (**I**).

Note hyphae emerging from the grafted site (**A**), around severely infected saplings (**B**, arrows), and rotted cortical tissue (**C**). *R. oryzae* was consistently isolated from rotted tissues of stocks of the saplings, and pathogenic to healthy stocks.

(Yoshida et al., 2001)

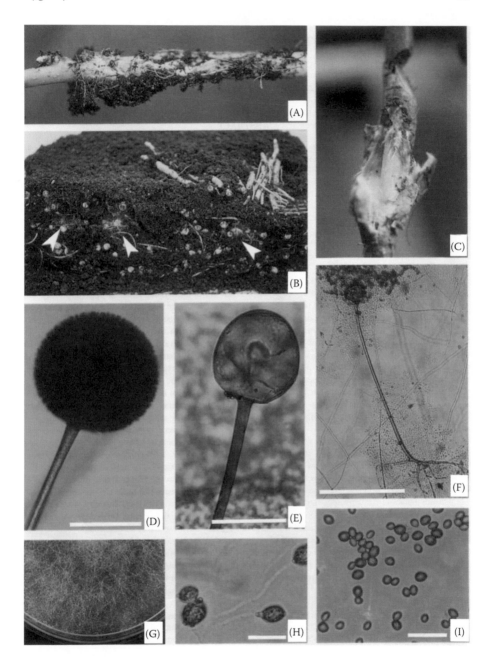

4.2 MORPHOLOGIES: *MUCOR HACHIJYOENSIS* AND *M. MEGUROENSE*

A. Two homothallic *Mucor*, *M. hachijyoensis* (left, isolate 70–1179 (ATCC201000)) and *M. meguroense* (right, isolate 92–268 (ATCC200999)) on seven-day-old PDA cultures in plates were isolated from pineapple field soil and from *Phellodendron amurense* seedlings, respectively in Japan.

B. Colony diameter (mm) of *Mucor hachijyoensis* (solid square) and *M. meguroense* (blank square) 24 hours after inoculation on PDA dishes at 13 different temperatures (Watanabe, 1994).

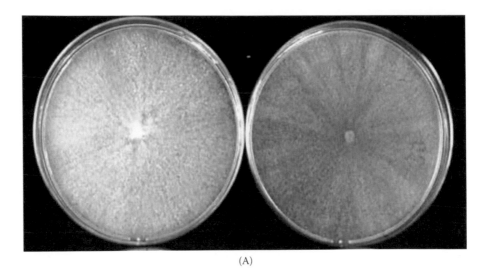

(A)

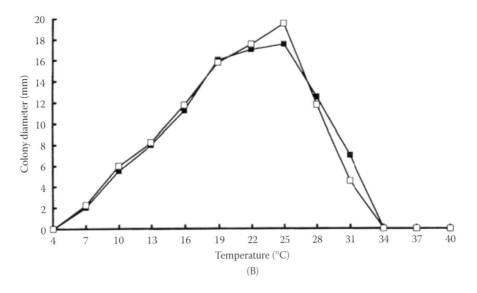

(B)

4.3 MORPHOLOGIES: *MUCOR HACHIJYOENSIS* AND *M. MEGUROENSE* (CONTINUED)

A. *Mucor hachijyoensis* (left) and *M. meguroense* (right) on seven-day-old PDA cultures in plates.

Mucor hachijyoensis

B. Part of sporangiophore and sporangium.

C–D. Part of sporangiophores with globose columellae, sporangiospores, and hyphae.

E. Sporangiospores.

F. Chlamydospores and hyphae.

G–J. Immature (**G, I**) and mature zygospores (**H, J**) with unequal suspensors.

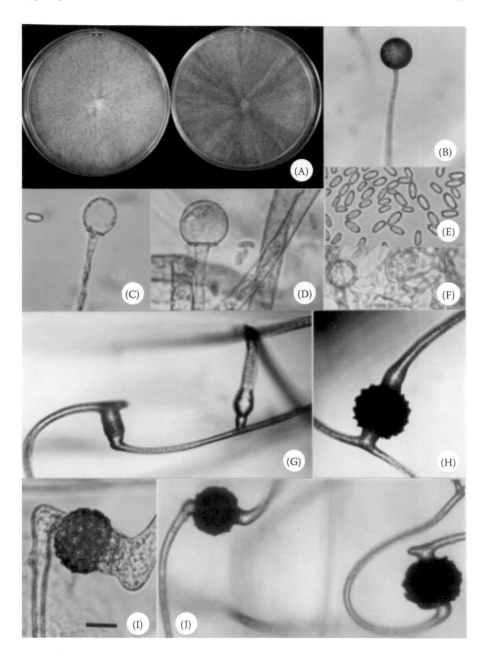

4.4   MORPHOLOGIES: *M. MEGUROENSE*

A. Part of sporangiophore, immature sporangium, sporangiospores, and hyphae.

B–D. Part of sporangiophores with clavate, elliptical, and cylindrical columellae and sporangiospores. Note distinct collars in Figures **C** and **D**.

E–G. Zygospores with unequal suspensors, and sporangiospores (**E**).

H. Chlamydospores.

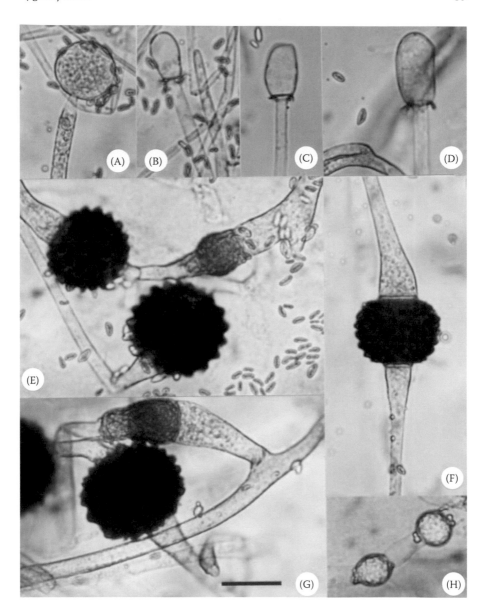

4.5 *MORTIERELLA TSUKUBAENSIS* ISOLATED FROM UNCULTIVATED SOIL IN TSUKUBA

A. Six-day-old PDA colony at 25°C in plate.

B. Variously-sized chlamydospores and hyphae.

C. Part of sporangiophore and sporangium.

D–E. Part of sporangiophores with columellae, sporangiospores and hyphae.

F. Sporangiospores.

G. Basal portions of common sporangiophore with vesicles rarely formed.

H. Immature zygospore with monoclinous major and minor suspensors.

I–J. Mature zygospores with major suspensors and a slightly inflated (I) or indiscernible or aborted minor suspensors (**J**).

The fungus, *M. tsukubaensis* (isolate 98-120 (ATCC204319)) is characterized by simple, large sporangiophores with terminal, many-spored sporangia, large chlamydospores, and naked zygospores. It is homothallic.

(Watanabe et al., 2001)

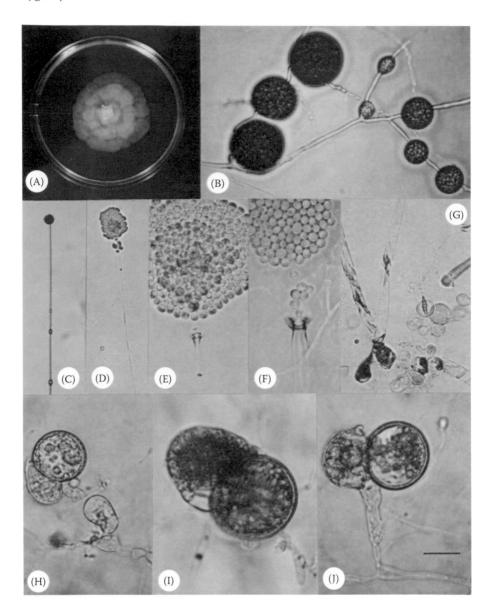

4.6   ECOLOGY: *MORTIERELLA* SPECIES FOUND IN JAPAN

Mortierella species were found not pathogenic to cucumber seedlings by the soil-over agar-culture method. Among a total of 73 *Pythium* and *Mortierella* isolates obtained from the Ogasawara Islands, none of the 21 *Mortierella* isolates and three of the uninoculated controls were pathogenic to cucumber seedlings; the seedlings that emerged were all healthy, even though some plates were covered with abundant mycelia. On the contrary, 39 out of 52 *Pythium* isolates were pathogenic, showing less than 30% healthy seedling stands.

Mortierella species were applicable to the diseased soil directly by sowing the artificially infested cucumber seeds. These *Mortierella* species were antagonistic against the damping-off pathogens of *Pythium* and *Rhizoctonia* species. For example, *M. alpina* and *M. tsukubaensis* were antagonistic against *P. sylvaticum*, *P. ultimum*, and *R. solani*. They were successfully applicable to the diseased soil by the soil-over agar-culture method just by sowing *Mortierella*-infected cucumber seeds.

Mortierella species from Hachijō-jima Island, Japan.

A. *M. elongata* (isolate 70–1074).

B. *Mortierella* (isolate 70–1388).

C. *Mortierella* (isolate 70–1373).

D. *Mortierella* (isolate 70–1673).

E. *Mortierella* (isolate 70–1480).

F. *M. alpina* (isolate 70–1394).

(Watanabe et al., 2000).

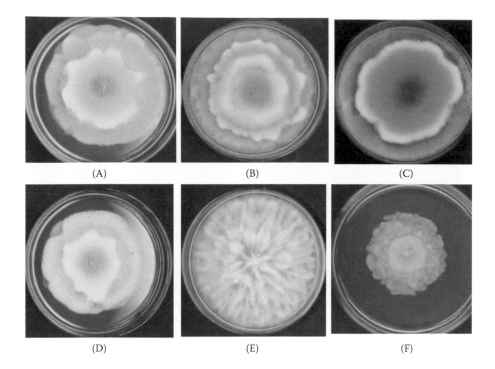

(A) (B) (C)

(D) (E) (F)

4.7 *MORTIERELLA ALPINA* AND *M. TSUKUBAENSIS* AS PROMISING BIOCONTROL AGENTS AGAINST DAMPING-OFF PATHOGENS OF CUCUMBER SEEDLINGS

A–D. Clear inhibition zones (**A, C**) formed by antagonistic *Mortierella alpina* (isolate 98-167), and *M. tsukubaensis* (isolate 98-120 (ATCC204319)) against five species each of *Pythium* (**B**) and *Rhizoctonia* (**D**).

A. Clear inhibition zone formed by *Mortierella alpina* (isolate 98–167).

B. Top row: *Pythium dissotocum* (isolate 98–66) and *P. sylvaticum* (isolate 98–117). Bottom row: *P. sylvaticum* (isolates 98–118 and 98–119).

C. Clear inhibition zone formed by *M. tsukubaensis* (isolate 98–120 (ATCC204319)).

D. Top row: *Rhizoctonia fragariae* (isolate 98–42) and *R. solani* (isolate 98–43). Bottom row: *R. solani* (isolate 98–44) and *Rhizoctonia* sp. (isolate 98–112).

E–F. Both *M. alpina* and *M. tsukubaensis* were effective in seed treatment for control of the damping-off in the diseased soils caused by *P. sylvaticum* (**E**, isolate 98–117) and *R. solani* (**F**, isolate 98–43, rightmost) and nontreated seed in center plates in relation to noninfested controls (leftmost).

Pythium dissotocum (isolate 98–66), *P. sylvaticum* (isolates 98–117, 98–118, and 98–119), and *Rhizoctonia fragagariae* (isolate 98–42), *R. solani* (isolate 98–43), *R. solani* (isolate 98–44), and *Rhizoctonia* sp. (isolate 98–112) are used as test fungi.

(A)

(B)

(C)

(D)

(E)

(F)

4.8 SIMPLE ASSAY ON SOILBORNE PLANT PATHOGENIC FUNGI IN OGASAWARA ISLANDS, JAPAN

Pathogenicity of soil fungi may be simply determined by the soil-over-agar culture inoculation method using 21 *Mortierella* and 52 *Pythium* strains isolated in the Ogasawara Islands.

The agar cultures in plates (9 cm in diameter) are covered with topping soil (50 mL/plate), the seeds are sown (placed on soil) (i.e., ten cucumber seeds), watered (10 mL), covered with additional topping soil (30 mL), and rewatered (10 mL).

None of 21 *Mortierella* isolates, and three uninoculated controls were pathogenic to cucumber seedlings, and the seedlings that emerged were all healthy, even though some plates were covered with abundant mycelia. On the contrary, 39 out of 52 *Pythium* spp. were pathogenic, with less than 30% healthy seedling stand rates.

TABLE 4.1

Pathogenicity of 21 *Mortierella* and 52 *Pythium* Isolates of the Ogasawara Islands by the Soil-Over-Agar Culture Inoculation Method

Healthy Stand (%)	*Pythium*	*Mortierella*	Control
0–10	26[a]	0	0
20–30	13	0	0
Over 40	13	21	3
Total	52	21	3

[a] Number of isolates belonged to the respective healthy stand rate.

(Watanabe et al., 2002)

A–C. No seedlings emerged from the *Pythium*-inoculated plates (**C**), compared to healthy seedlings in noninfested control (**A**), and seedlings inoculated with *Mortierella* (**B**).

Note the abundant mycelia from *Mortierella* inocula (**B**).

(A)   (B)   (C)

5 Ascomycetes

5.1   DISEASES: ANTHRACNOSE

A. Anthracnose of pinto bean at Nagano.

B. Damping-off of pinto bean seedlings caused by seedborne anthracnose pathogen, *Colletotrichum lindemuthianum* (anamorph).

(A)

(B)

5.2   MORPHOLOGIES: ANTHRACNOSE PATHOGEN, *GLOMERELLA GLYCINES* (TELEOMORPH)

A. Crushed perithecium and discharged asci.

B. Clustered asci bearing ascospores ejected from crushed perithecium.

C. Curved elliptical ascospores.

D. Conidiomata with setae and cylindrical conidia (isolate 83–189).

(*G. glycines* (teleomorph): perithecia 200–240 µm in diameter; asci 80–90 × 9.5–10.5 µm; ascospores 23.7–27.5 × 4.5–5 µm. *Colletotrichum* (anamorph): sporodochia 250–260 µm in diameter; setae 130–170 × 5–5.3 µm. Conidia 20 × 4.7–5.3 µm)

(Kurata, 1960; Tiffany and Gilman, 1954)

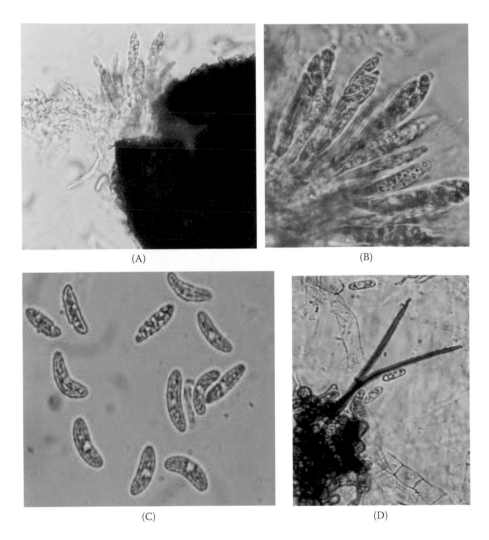

(A)

(B)

(C)

(D)

5.3   MORPHOLOGIES: ANTHRACNOSE PATHOGEN, *COLLETOTRICHUM* SPECIES (ANAMORPH)

A. *Colletotrichum* species from carrot leaves (isolate 93–325).

B. Acervuli and setae from *Phellodendron* isolate (93–337).

C. Conidiophores with conidia.

D. Conidia.

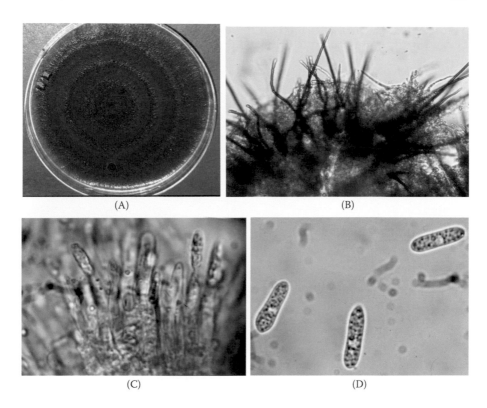

(A)

(B)

(C)

(D)

5.4   DISEASES: *BAKANAE* ("FOOLISH SEEDLING") DISEASE OF RICE PLANTS

A. Healthy (left pot), and diseased elongated seedlings (center and right).

B. Diseased seedlings with basal necrotic lesions.

(Watanabe and Umehara, 1977)

(A)   (B)

5.5 MORPHOLOGIES: *GIBBERELLA FUJIKUROI* (TELEOMORPH) AND *FUSARIUM MONILIFORME* (ANAMORPH)

A. Habit of perithecia of *Gibberella fujikuroi* (teleomorph) on rice culm.

B–D. Perithecia and pseudoparenchymatus perithecial wall (**D**).

E. Ascospores inside an ascus.

F. One- and two-septate ascospores.

G–I. Macro- (**I**) and microconidia (**G, H**) of *Fusarium moniliforme* (anamorph).

(*G. fujikuroi* (teleomorph): perithecia 148.2–287.5 × 172.9–345.8 µm; asci 50–105 × 4.5–8 µm; ascospores 10.5–19 × 3.5–7 µm: *F. moniliforme* (anamorph): microconidia 5–13 × 2.3–3 µm; macroconidia 42.5–62.5 × 2.5–3 µm).

(Watanabe and Umehara, 1977)

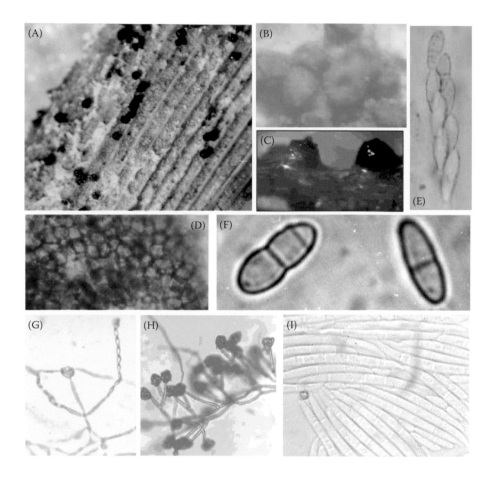

5.6   DISEASES: MELON ROOT ROTS

Melon root rots caused by *Monosporascus cannonballus*

A–B. Occurred at greenhouse in Yamagata (**A**) and Chiba (**B**).

C. In field in Chiba.

(Watanabe, 1979)

(A)

(B)

(C)

5.7   MORPHOLOGIES: MELON ROOT ROT PATHOGEN, *MONOSPORASCUS CANNONBALLUS*

A–B. Melon root rots with black dots showing perithecia.

C. Asci and ascospores ejected from crushed perithecium.

D. Ascus bearing one *cannonballus*-like ascospore.

(*Monosporascus cannonballus* (isolate 78–2494): perithecia 222–268 μm in diameter; asci 50–110 × 35–50 μm. Ascospores 32–47.5 μm in diameter)

(Watanabe, 1979)

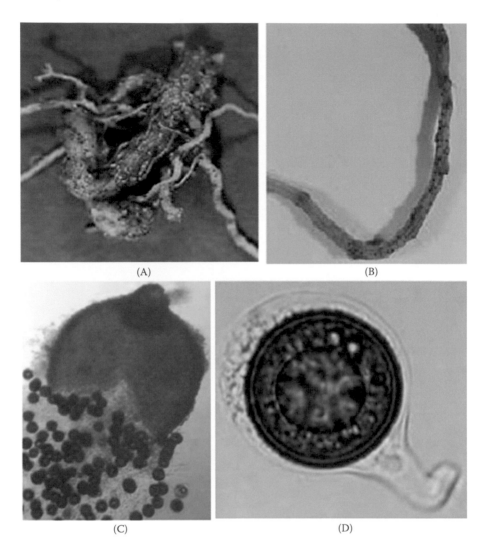

(A)

(B)

(C)

(D)

5.8 MORPHOLOGIES: *NECTRIA ASAKAWAENSIS*

Nectria asakawaensis associated with *Armillaria meria* rhizomorphs

A. Crushed perithecium, and ascospores released.

B. Perithecium wall and ascospores.

C. Asci and ascospores.

(*Nectria asakawaensis*: perithecia 175–310 × 140–240 µm; asci 75–87.5 × 15–20 µm; ascospores 18.7–25 × 10–12.5 µm)

The fungus (isolate 84–133 (MAFF240308)) has been tested as a biocontrol agent against soilborne plant pathogens.

(Watanabe, 1990)

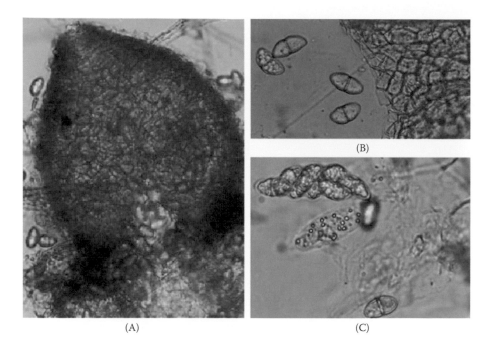

(A)

(B)

(C)

6 Basidiomycetous Fungi

6.1 DISEASES: ROOT ROTS OF APPLE AND CHERRY TREES

Root rots of apple, and flowering cherry trees caused by *Armillaria* spp.

A. Damaged apple tree infected with *A. mellea* in California, USA.

B. Flowering cherry tree infected with *Armillaria* sp. in Hachiōji, Japan.

C–D. Basidiocarps of *A. mellea* formed basally at flowering cherry tree in Tsukuba, Japan. Note the rhizomorph elongating basally from the basidiocarps.

(A)   (B)

(C)   (D)

6.2   MORPHOLOGIES: *ARMILLARIA MELLEA, A. TABESCENS*

Armillaria mellea and *A. tabescens*

A. *A. mellea* basidiocarps with annuls (the ring around the stem), which were not observable in this photo.

B. *A. tabescens* basidiocarps without annuls.

C–D. Basidia (**C**) and basidiospores (**D**) of *A. mellea*.

(Watanabe, 1986)

(A)

(B)

(C)

(D)

6.3   MORPHOLOGIES: *A. MELLEA* RHIZOMORPHS

A. Rhizomorphs associated with rotted root wood tissues.

B. White hyphal membrane under epidermal tissue.

C–D. Elongated shoestring-like rhizomorphs (**C**), and rhizomorph segments (**D**).

E–F. Aggregated slender hyphae elongated around the intermedium of rhizomorph (**E**), and the dissected rhizomorph (**F**).

G. Part of hyphae elongated from rhizomorph tissue on agar.

H. Rhizomorphs grown in potato dextrose broth in tubes.

(Watanabe, 1986)

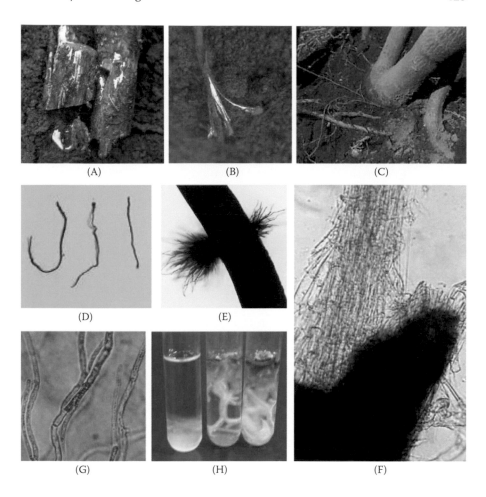

(A)

(B)

(C)

(D)

(E)

(G)

(H)

(F)

6.4 TEMPERATURE RESPONSES OF *A. MELLEA* AND *A. TABESCENS*

A. Temperature responses of *A. mellea* (isolate 84–128 (MAFF425282)) and *A. tabescens* (isolate 86–17 (MAFF425286)) on potato dextrose agar. The optimum temperature for mycelial growth was 28°C for both species. However, *A. mellea* favored higher temperatures than *A. tabescens*.

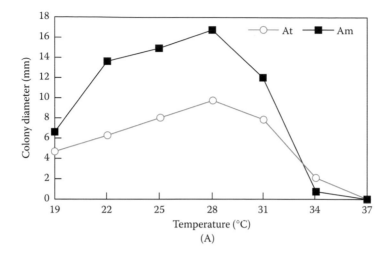

(A)

7 Deuteromycetous (Anamorphic) Fungi

7.1   DISEASES: *PHELLODENDRON AMURENSE* DECLINE

Phellodendron amurense is one component of the natural flora in the twenty-hectare garden of The Institute for Nature Study, Natural Museum in Tokyo, Japan. Among a total of 155 mature trees (9–76 cm in diameter at breast height) in the garden, 36 trees were dead and nearly 30 trees were diseased by the end of December 1993. Diseased trees were characterized by loss of turgidity and delayed new leaflet development in the spring. Leaflets that developed on the diseased trees were smaller than those on healthy ones. Canopies of the diseased trees were less dense than those of healthy trees because of less numerous and smaller leaflets (Figure A, p. 129). As disease progressed, new leaflets did not develop in the spring, and dieback of twigs occurred, resulting in the decline of the entire trees. Diseased trees had damaged root systems. The number of the lateral roots was reduced, and there increased the numerous discolored roots. The root epidermis did not appear healthy, was readily sloughed off, water-soaked, and became dark yellow. Abundant seedlings emerged near mature trees but died within a year. Diseased seedlings were often brown, had smaller roots and basal stems, and the root epidermis was disintegrated or sloughed-off. Diseased leaflets had slightly sunken, brown, necrotic lesions with indistinct margins. Three *Cylindrocladium* species were isolated from the diseased seedlings together with the fungi belonging to 24 fungus genera in 1992.

In inoculation tests, *C. colhounii* caused leaf blight and subsequent abscission of leaflets of seedlings, saplings, and trees, and damping-off of potted seedlings, but *C. camellea* and *C. tenue* were not pathogenic. Morphologies of the pathogens and their temperature responses are described subsequently.

(Watanabe, 1994; Watanabe et al., 1995)

A. Declining *Phellodendron* tree (right) in comparison with healthy tree (left).

B. Symptoms on a leaflet of a mature tree (left) four days after inoculation with *Cylindrocladium colhounii* (isolate 92–211 (ATCC201116)) and a healthy leaflet (right) with PDA disk (arrow) as control following removal of wetting material.

C. Blighted leaflet (left) inoculated with *C. colhounii* and healthy leaflet used as control (right) together with inocula (agar disks) on pieces of cotton cover to keep inocula moist.

D. Seedling stand collapsed in the infested soil with *C. colhounii* (right) and healthy stand in noninfested soil (left).

E. The collapsed seedling with discolored basal lesion.

(Watanabe et al., 1995)

(A)

(B)

(C)

(D)

(E)

7.2 MORPHOLOGIES: *CYLINDROCLADIUM* SPECIES

A. Sporulation of *Cylindrocladium* species associated with *Phellodendron amurense* trees.

B. Conidiophores with stipe and a portion of vesicle (arrow) and three septate conidia of *colhounii*.

C. Seven-day-old PDA colonies of leaf (left, isolate 92–260 (MAFF425361)), and root isolates (right, isolate 92–211 (ATCC201116)) of *C. colhounii*.

D. Seven-day-old PDA colonies of *C. camelliae* (left, isolate 92–202 (ATCC201118)) and *C. tenue* (right, isolate 92–246 (ATCC201311)).

(*C. colhounii*: conidiophores branched once or twice, branches with phialides and spore masses, together with stipes and terminal clavate or ellipsoidal vesicles, conidia phialosporous, four-celled, reddish brown, catenulate or aggregated chlamydospores, and conidiophores including stipes and terminal vesicles 215–475 × 6–10 μm; phialides 7.5–17.5 × 2.5–6.2 μm; vesicles 12.5–65 × 2–4.5 μm; conidia 45–77.5 × 4.7–6 μm; chlamydospores 15–20 μm in diameter.)

(Watanabe, 1994)

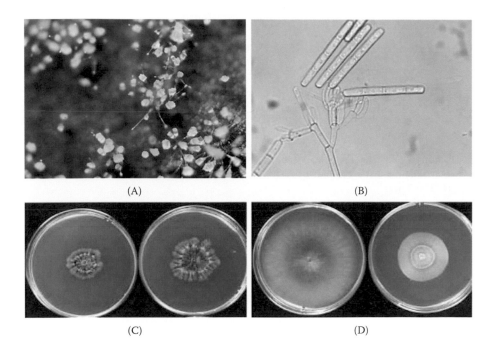

(A)

(B)

(C)

(D)

7.3   MORPHOLOGIES: *C. TENUE*

Cylindrocladium tenue ((isolate 92–246 (ATCC201311)).

A–D. Habit showing *Gliocladium*-like state.

E. Conidiophore with verticillate phialides and a conidium.

F. Conidiophore with branches and terminal phialides bearing two immature conidia.

G. One-septate conidia.

H–J. Conidiophore branches and terminal phialides, bearing single (**I**) and several (**H**) immature conidia. Scale bar in (**I**): for (**A–B**) = 35 μm; for (**C–F**) = 18 μm; for (**G**) = 13 μm; for (**H–J**) = 7 μm.

(Watanabe, 1994)

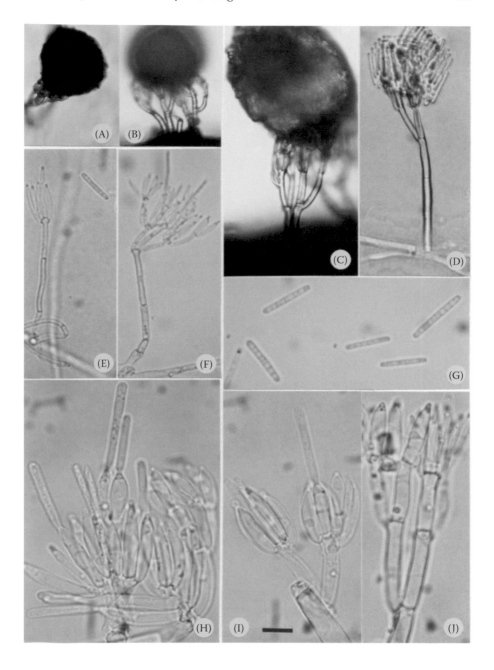

7.4   MORPHOLOGIES: DRIED SPECIMENS
OF *CYLINDROCLADIUM* SPECIES

A. *C. camelliae* (isolate 92–202 (ATCC201118)).

B. *C. citri* (isolate 00–52 (MAFF238171)).

C. *C. floridanum* (isolate 02–224).

D. *Cylindrocladium* sp. (isolate 92–111).

E. *C. scoparium* (isolate 92–118 (NBRC32535)).

F. *C. tenue* (isolate 02–215).

G. *C. camelliae* (isolate 01–204).

H. *C. colhounii* (isolate 92–260 (MAFF425361)). This sample is older than eight days.

I. *C. colhounii* (isolate 92–260 (NBRC32530)). Eight-day-old sample.

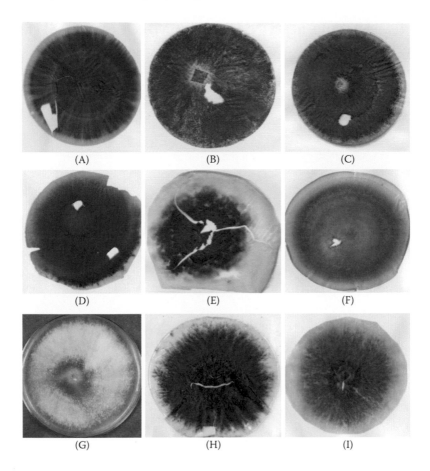

7.5 TEMPERATURE RESPONSES OF *CYLINDROCLADIUM* SPECIES

A. Colony diameter (mm) of four isolates of *Cylindrocladium* species: *C. camelliae* (isolate 92–202 (ATCC201118): large square), *C. colhounii* (isolate 92–211 (ATCC201116): circle, isolate 92–260 (NBRC32530): triangle), and *C. tenue* (isolate 92–246 (ATCC201311): small square) after incubation for five days on PDA at 11 different temperatures.

(Watanabe, 1994)

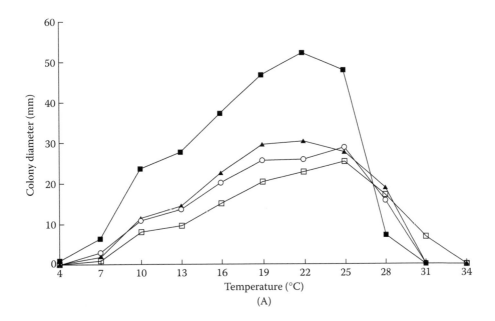

(A)

7.6   DISEASES: TURNIP YELLOWS AND BIOCONTROL PRACTICES

Turnip yellows were first reported in Japan in 1934; since occurrence of the disease in Kashiwa, Chiba in 1993, biocontrol practices have been attempted.

The diseased plants were poor in growth, stunted, and wilted; they turned yellow and finally collapsed. Vascular bundles of the diseased plants were discolored.

Fusarium oxysporum was readily isolated from the discolored vascular tissues, and its pathogenicity was demonstrated by the discolored bulb sections inoculated by *F. oxysporum* agar discs placed onto the clean healthy sections. The fungus (isolate 02–337 (MAFF240304)) was identified as *Fusarium oxysporum* f. sp. *raphani*.

Several *F. oxysporum* isolates and soil fungi from the turnip field and the rhizosphere soils were also tested for the pathogenicity, but they did not cause any damage, although some inoculated sections were slightly discolored because of their morphological structures, that is, perithesia of *Chaetomium spirale*, or dark conidia of *Cladorhinum* sp.

Some isolates were antagonistic against *F. oxysporum* f. sp. *raphani* in dual cultures on PDA.

The emergence rate of healthy seedlings grown in the diseased soil was 44%, whereas it was over 70% in the soil treated with *Gliocladium* sp., *Myrothecium* sp., and *Papulaspora* sp.

TABLE 7.1
Major fungi from turnip field soils at Kashiwa, and used as biocontrol agents

Alternaria alternata (isolate 02–350)	*Aspergillus fumigata* (02–372)
Chaetomium spirale (02–359)	*Cladorhinum* sp. (02–354)
Colletotrichum sp. (02–371)	*Cylindrocarpon* sp. (02–367)
Fusarium oxysporum (02–369, 02–361)	*F. roseum* (02–358, 02–363)
F. solani (02–353)	*Fusarium* sp. (02–348)
Gliocladium sp. (02–366)	*Humicola* sp. (02–362)
Myrothecium verrucaria (02–365)	*Paecilomyces* sp. (02–374)
Papulaspora sp. (02–365)	*Pythium spinosum* (02–345)
P. sylvaticum (02–344)	Unknowns (02–356, 02–376)

Source: Watanabe et al., 2003

7.7   DISEASES: TURNIP YELLOWS

A. Three bulbs each of diseased (left) and healthy turnips (right) from the naturally infested field at Chiba 43 days after sowing.

B. Two diseased half-cut bulbs (left, center) and one healthy half-cut bulb (right).

C. Turnip bulbs inoculated with *Cladorhinum* sp. (top left), *Chaetomium spinosum* (top right), uninoculated (lower left), and *F. oxysporum* (lower right).

D–F. Discolored turnip bulb inoculated with *F. oxysporum* (**D**), dotted (perithecia) with *Chaetomium spirale* (**E**) and stained with *Cladorhinum* sp. ten days after inoculation (**F**).

(Watanabe et al., 2003)

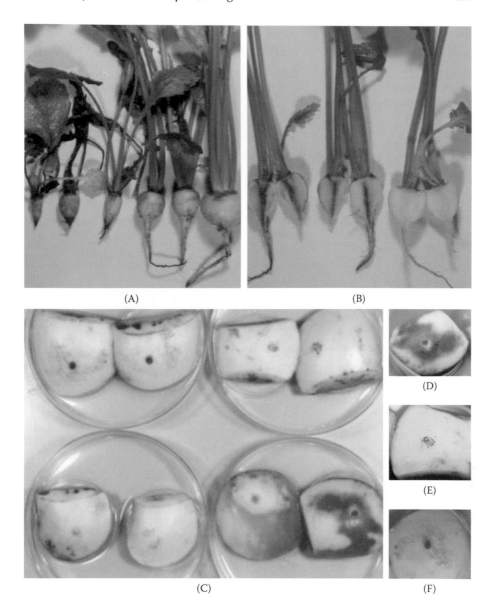

(A)

(B)

(C)

(D)

(E)

(F)

7.8   MORPHOLOGIES: TURNIP YELLOWS PATHOGEN, *FUSARIUM OXYSPORUM* F. SP. *RAPHANI*

A–B. Pathogenic isolate (02–337 (MAFF240304)) (**A**), and two nonpathogenic isolates of *F. oxysporum* (**B**).

C. Macro- and microconidia.

D–E. Spore masses composed of microconidia (**D–E**) and chlamydospores (**E**).

(*F. oxysporum*: macroconidia 26–33 × 5–6 μm; microconidia 6–20 × 4–5 μm; chlamydospores 6–16 μm in diameter)

(Watanabe et al., 2003)

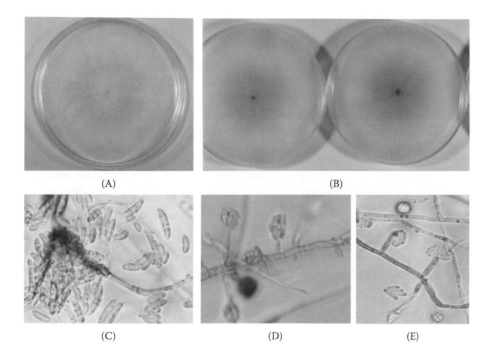

(A)

(B)

(C)

(D)

(E)

7.9 BIOCONTROL PROCEDURE FOR TURNIP YELLOWS WITH ANTAGONISTIC FUNGI

A, B. Coculture of *F. oxysporum* f. sp. *raphani* (center) with *Aspergillus fumigatus* (top) and *Paecilomyces* sp. (bottom) at the periphery of the plate forming inhibition zones between them. Picture taken after eighteen days of culture.

C. Damping-off of turnip seedlings by soil treatment of antagonistic fungi showing healthy seedlings grown in autoclaved naturally-infested soil (lower left), damped-off seedlings in naturally-infested soil (lower center), and naturally-infested soil infested with *F. oxysporum* f. sp. *raphani* (lower right), and rather healthy seedlings grown in soils (top four plates) treated with antagonistic fungi, that is, *Papaulaspora* sp. (isolate 02–335), *Chaetomium spirale* (isolate 02–259), *Gliocladium* sp. (isolate 02–366), and *Myrothecium* sp. (isolate 02–376) from top left to top right. Picture was taken 12 days after sowing.

(Watanabe et al., 2003)

7.10 DISEASES: SOIL- AND SEEDBORNE PLANT PATHOGENS

Some local commercial kidney bean seeds were poor in emergence and seedlings diseased after emergence. In surveying a total of 1036 fungal isolates from the 6390 seeds, 24 genera were identified. *Alternaria, Fusarium, Colletotrichum, Chaetomium* and *Rhizoctonia* were frequently isolated and they occupied 86% of the total isolates. Kidney bean plants became diseased in the autoclaved soil artificially infested with *Colletotrichum lindemuthianum, Rhizoctonia solani* or *Macrophomina phaseolina* isolated from these seeds. One per 22–41 commercial seeds were infected with at least one of these three fungi.

Charcoal rot pathogen, *Macrophomina phaseolina*, has been reported in Japan since 1942, but the disease incidence has not been severe, although the diseases had been significant in tropical and subtropical areas, and over 300 plant host species have been reported worldwide. The local commercial kidney bean seeds (var. Topcrop) were assayed for the fungus, and in 280 visually healthy seeds examined, 2.1% of the seeds were infected by this fungus. An average of six viable microsclerotia of *M. phaseolina* were detected in the contaminated dirt per 100 grams of these seeds. The field soil of the seed producing area was assessed to contain an average of 11 viable microsclerotia per gram of air-dried soil, and kidney plants grown in the soil became infected under the green house conditions.

Macrophomina phaseolina (isolate 68–4 (MAFF305189)) is morphologically characterized by the tiny sclerotia (nearly 100 μm in diameter) and pycnidia (130.0–250.0 × 115.0–230.0 μm) with hyaline large elliptical pycnospores (14.0–35.0 × 6.0–11.5 μm). Ostioles were 15.0–35.0 μm in diameter.

An average of nine propagules of the fungus per gram of soil were initially detected, but after 2.5 or 4.17 years' storage, the number was reduced to four and two propagules per gram of soil, respectively. At two different sugar cane field soils in Okinawa, three or eight propagules were detected after 3.5 years' storage, but after 5.75–6 years, the fungus was never detected (Figures A, B on p. 153).

(A)   (B)

(C)

On the assay of longevity, cultured sclerotia of three tested isolates were 16%–40% viable, and in four years, they were 5%–8% viable. In the tissue-formed sclerotia, four isolates were 1%–27% viable after three years, and the soil-formed sclerotia originated from water agar-and PDA were 4% and 56% viable after three years, respectively (Watanabe, 1972, 1973).

7.11 DISEASES: CHARCOAL ROT OF PINTO BEAN AND *EUCALYPTUS* SEEDLINGS

A. Healthy (left) and diseased (right) kidney bean seeds infected with *Macrophomina phaseolina*.

B. Charcoal rot of *Eucalyptus* seedlings.

C. Charcoal rot of pinto bean hypocotyl.

TABLE 7.2

Microsclerotia of *M. phaseolina* Contaminated in the Plant Residue and Dirt Washed Off from Three Varieties of Commercial Kidney Bean Seeds

| Variety | Weight | | No. of Microsclerotium-like particles | No. of Viable Microsclerotia |
	Seed	Dirt		
Topcrop	100 g	16 mg	16	4
	500	220	51	31
Edogawa-tsurunashi	100	18	19	0
	500	120	84	0
Tsurunashi-shakugosun	100	12	7	0
	500	140	34	0

(Watanabe, 1972)

The fungus forms tiny sclerotia (microsclerotia), measuring around 100 μm in diameter in plant tissues, and they survive in soil. Among three commercial kidney bean variety seeds checked, only "Topcrop" was infected with *M. phaseolina* at a rate of 2.1%. The seed coat appeared to be infected because no fungus was recovered from the seeds without a seed coat (= peeled seeds).

(A)

(B)   (C)

7.12   DISEASES: CHARCOAL ROT OF PINTO BEAN

Macrophomina phaseolina microsclerotia recovered from the snap bean fields and their survivability (Watanabe, 1973).

A. *M. phaseolina*-infested fields in Nagano.

B. Tissue-formed microsclerotia.

C. Microsclerotia detected from the field soil.

D. Microsclerotia recovered from the field soil (top), PD broth-cultured (lower left) and tissue-formed (lower right).

E–F. *M. phaseolina* (isolate 68–4 (MAFF305159)): PDA colonies (**E**) and the dried specimen (**F**).

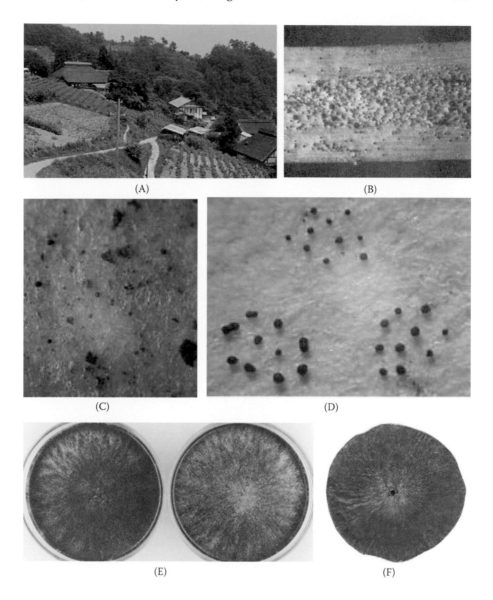

(A)

(B)

(C)

(D)

(E)

(F)

7.13   DISEASES: CHARCOAL ROT OF FOREST TREES AND PINTO BEAN SEEDLINGS

A–B. *Macrophomina phaseolina* associated with the decline of forest tree seedlings at the Nursery forest at CEDEFO in Paraguay.

C. Pathogenicity of *M. phaseolina* shown in potted pinto been seedlings 19 days after sowing. Basal and stem blights occurred on the emerged seedlings.

(Watanabe, 1973)

(A)   (B)

(C)

7.14 MORPHOLOGIES: MICROSCLEROTIA AND PYCNIDIA OF *MACROPHOMINA PHASEOLINA*

A. Microsclerotia and pycnidia (P) of *M. phaseolina* (isolate 70–1028 (MAFF425254)) formed on dried kidney bean hypocotyl segment seven days after inoculation.

B. Pycnidia taken out from the kidney bean segment.

C. Ostiolate pycnidia.

D. Pycnospores released from artificially ruptured pycnidia.

E. Pycnospores.

F. Pycnidiophores from a crushed pycnidium.

G–I. Germination of pycnospores seven hours after seeding on water agar at 26°C.

(Watanabe, 1972)

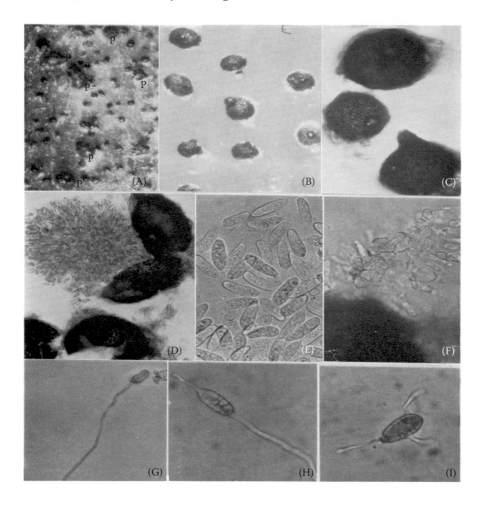

7.15 *M. PHASEOLINA* MICROSCLEROTIA POPULATIONS (NUMBER OF PROPAGULES PER GRAM OF SOIL)

A–B. Microsclerotia populations (number of propagules per gram of soil) of *Macrophomina phaseolina* in progression of storage length of soils from Okinawa (**A**) and Nagano (**B**).

(Watanabe, 1973)

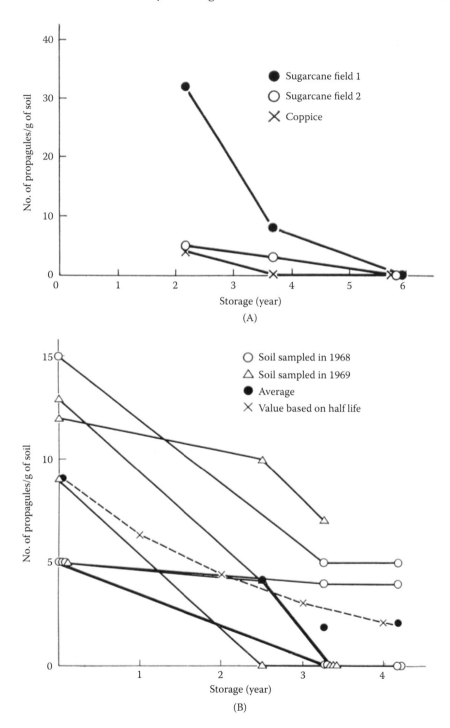

(A)

(B)

7.16 WHITE ROOT ROT PATHOGEN, *DEMATOPHORA NECATRIX* (ANAMORPH)

White root rots of various plants and their causal fungus, *Rosellina necatrix*, have been known in Europe as early as the 1890s, and in Japan in 1902 (Araki, 1967). Both teleomorph and anamorph morphologies of the fungus have been illustrated worldwide. Although there is numerous literature on the diseases and *R. necatrix,* there have been scarce reports available on the sporulation in vitro.

The fungus (isolates 84–473 (MAFF425314); 84–374 (MAFF425317)) was isolated from collapsed poplar tree root tissues in Ehime, Shikoku. It formed white mycelia with pear-shaped swellings near septa in hyphae characteristically (Section 7.18, Figure A, p. 157).

Pathogenicity of the fungus was tested by the soil-over-agar culture inoculation method (Table 7.2). In this method, seven-day-old PDA cultures were covered with autoclaved soil (50 mL/small 15-cm-deep plate or 100 mL/large 20-cm-deep plate) and watered appropriately whenever necessary, and kept moist. The noncultured PDA plates were similarly treated, and served as controls. Experiments were conducted in a growth chamber regulated at 25°C for 12 hours of light and 19°C for 12 hours of darkness.

Pathogenicity was evaluated by rates of emergence, and damaged seedlings (collapsed or diseased seedlings)/total number of seeds tested 7–40 days after sowing.

Isolate 84–374 was more pathogenic than isolate 84–373 (Table 7.3). Namely 26%–100% of the seedlings were diseased or collapsed by isolate 84–374, whereas 3.3%–46% were by isolate 84–373. The seedlings in the noninfested soil as a control were healthy. One mutated isolate, 84–373 (O), 84–373 (P) and (D), was weakly pathogenic, causing 4%–6% damage rates. The diseased plants became discolored and narrowed basally near the soil surface. White mycelia covered all over the soil surface, or the collapsed seedlings.

TABLE 7.3

Pathogenicity of Two *Dematophora Necatrix* Isolates to Black Pine Seedlings by the Soil-Over-Agar Culture Inoculation Method

Treatment	Seed (no.)	Emergence (%)	Healthy Stand
Non-inf. control	130	85.3 (76–90)	85.3 (76–90)
Isolate 84–373	130	72.7 (60–88)	61.2 (45–72)
Isolate 84–374	50	74.0	0

(Watanabe, 1992)

7.17 PATHOGENICITY OF WHITE ROOT ROT PATHOGEN, *DEMATOPHORA NECATRIX* TO BLACK PINE SEEDLINGS

A–B. Black pine seedlings stand in noninfested (left) and infested soil (right) with *Dematophora necatrix* (isolate 84–374 (MAFF425317)) in side (**A**) and top view (**B**).

(Watanabe, 1992)

(A)

(B)

(C)

(previous page)

C. Black pine seedling stand in noninfested (left, two dishes) and infested soil (right) with
D. necatrix (isolate 84–373 mutated "W" & "D," center two dishes) and *D. necatrix* (isolate
84–373 "P," top right) and isolate 84–374 (below right).

(Watanabe, 1986, 1992)

7.18   MORPHOLOGIES: WHITE ROOT ROT PATHOGEN, *DEMATOPHORA NECATRIX*

A. Pear-shaped swellings near hyphal septa (arrows).

B–C. Normal (**B**) and mutated PDA colonies of *D. necatrix* (**C**).

D. Conidiophores and conidia.

E. Apical conidiogenous cells showing crater-like conidial scars after conidial detachment.

F. Germinated conidia.

G. PDA colony diameter (mm) of four isolates (84–373 (O): triangle, 84–373 (D): square,
84–373 (P): dark circle, 84–374: blank circle) of *D. necatrix* after incubation for five days at
13 different temperatures.

(*D. necatrix* (anamorph): conidiophores more than 500 μm tall, 2.5–2.8 μm wide, bearing
2–30 conidia in two rows. Conidia 3.7–5 × 2–2.2 μm).

(Watanabe, 1992)

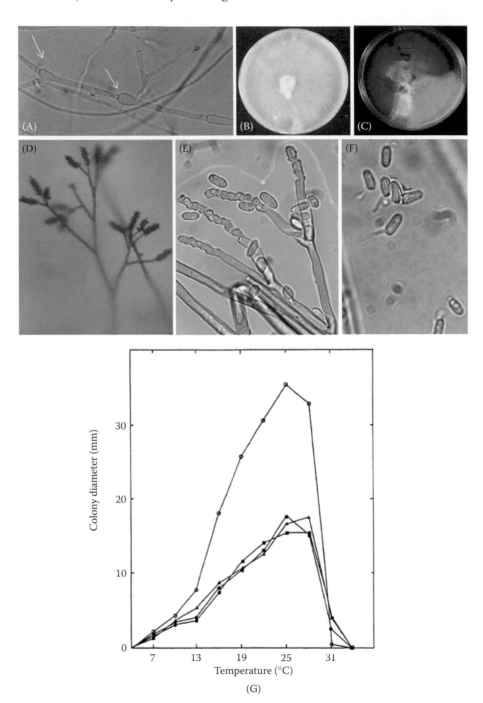

(G)

7.19 DISEASE AND MORPHOLOGIES: ROOT ROT OF MELON CAUSED BY *NODULISPORIUM MELONIS*

Melon plants were collapsed with brown-rotted and corky roots in Shizuoka, Japan in 1975. A new *Nodulisporium* species, *N. melonis* (isolate 73–355 (ATCC200606)) was predominantly isolated from the diseased tissues with five other fungus genera. In inoculation tests, *N. melonis* was pathogenic, but others were not. PDA colonies of the fungus are white and homogeneous, sub-hyaline, cream to pale brown in reverse.

The conidiophores are hyaline, simple or branched verticillately or irregularly, 45–280 × 1.2–2.3 μm, bearing single or apical spore masses, 2–9 conidia per head on apical fertile portions, denticulate after detachment of conidia, often proliferate from conidia on conidiophores. Conidia are single, in short chains or aggregated, sympodulosporous, acropleurogenous, hyaline, single-celled, obovoid, cylindrical, elliptical or irregular in shape, 2.5–13.8 × 1.2–3.0 μm with basal frill or cylindrical appendix, 3.0–10.0 × 0.4–1.0 μm.

(Sato et al., 1995, Watanabe and Sato, 1995)

7.20 DISEASES: MELON ROOT ROT

Root rot of melon caused by *Nodulisporium melonis*.

A–B. Damaged roots showing typical brown corky root rot after disappearance of secondary roots and rootlets.

C. Three plants each grown in the artificially infested soils (right), and in the noninfested soil (left). Note the damaged roots in the infested soil.

(Sato et al., 1995).

(A)   (B)

(C)

7.21 MORPHOLOGIES: *NODULISPORIUM MELONIS*

A–E. Conidiophores bear apical spore masses.

F. Conidia germination directly on the spore masses.

G. Detached conidia.

H. Two conidia at the apices of conidiophore.

I. Conidium on the apical phialide.

J. Conidium removed from apical phialide.

K. Elongated conidia at the apex.

(Watanabe and Sato, 1995)

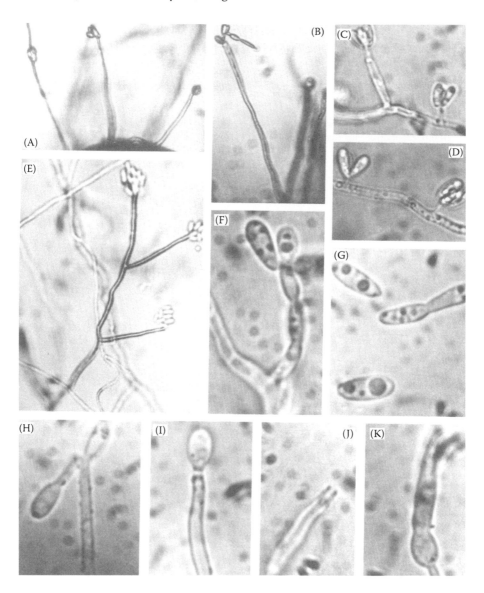

7.22   MORPHOLOGIES *NODULISPORIUM MELONIS,* TEMPERATURE RESPONSE, AND SPECIMEN

A. Colony diameter (mm) of two isolates of *Nodulisporium melonis* (isolate N–1, 75–355 (ATCC200606): triangle; isolate N–8, 75–356: circle) seven days after inoculation on PDA at 12 different temperatures.

B. Dry culture specimen of *Nodulisporium melonis* (isolate 75–356 (ATCC200607)). The optimum temperature for mycelial growth on PDA was 24°C.

(Watanabe and Sato, 1995)

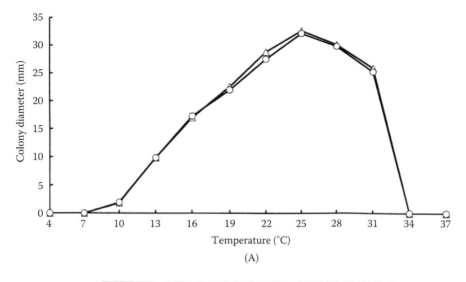

(A)

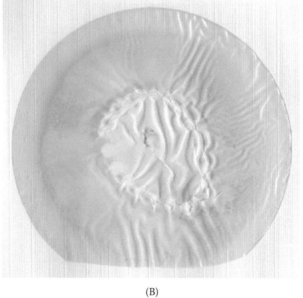

(B)

7.23   DISEASES: GENTIAN PINK ROOT ROT

A. Healthy gentian plants at the field.

B. Pink root rot of gentian caused by *Pyrenochaeta* species.

C. Internally discolored root in the tangentially cut surface.

D. Shriveled and discolored root tip.

(A)

(B)

(C)

(D)

7.24 DISEASES: GENTIAN PINK ROOT ROT PATHOGENS, INOCULATION

A–B. Symptoms developed on the right three gentian shoots (A) and two right root segments (B) 12 days after inoculation with *P. gentianicola* compared with healthy segments with sterile PDA discs as controls.

TABLE 7.4

Pathogenicity[a] of *P. gentianicola* (Isolate 76–501 (ATCC200774)) and *P. terrestris* (Isolate 76–502 (ATCC200773)) to Gentian Root Segments 12 Days After Inoculation at 16°C, 20°C, and 26°C

	Diseased Segments (%)			Lesion Length (mm)		
	16°C	20°C	26°C	16°C	20°C	26°C
Control	0	28	16.7	0	0	0.1
P. gentianicola	70.9	80.6	54.2	6.6	6.3	4.9
P. terrestris	4.2	38.9	45.9	0	2.3	7.8

[a]   Percentage of diseased root segments in a total of 18 roots inoculated in each treatment, and average lesion length per segment 12 days after inoculation in three separate experiments.

(Watanabe and Imamura, 1995)

7.25   MORPHOLOGIES: GENTIAN PINK ROOT ROT PATHOGENS

Cultivated gentian (*Gentiana scabra* var. *buergeri*) has been seriously damaged by pink root rot since the 1970s. Diseased plants finally collapsed after wilting and stunting. Among over 40 fungal genera isolated, both *P. gentianicola* (isolate 76–501 (ATCC200774)) and *P. terrestris* were (isolate 76–502 (ATCC200773)) always associated with the diseased plants. Particularly, *P. gentianicola* was isolated for the first time, and named as a new species and shown as one of the pathogens of this disease.

The pathogenicity of both species was demonstrated (Table 7.4, p. 166).

A. PDA colonies of *P. gentianicola* (top three plates) and *P. terrestris* (lower three plates) at 25°C (left), 28°C (center), and 31°C (right) 18 days after inoculation in 9-cm Petri dishes.

B. Colony diameter (mm) of *P. gentianicola* (= P. g.) and *P. terrestris* (= P. t.) five days after inoculation on PDA dishes at seven different temperatures.

(Watanabe and Imamura, 1995)

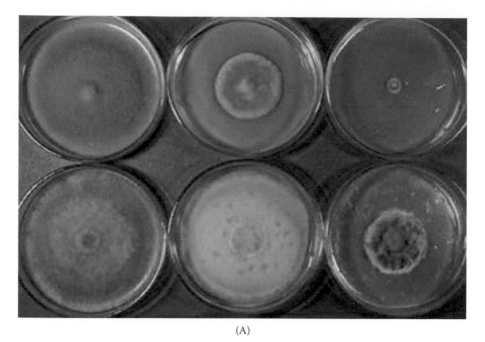

(A)

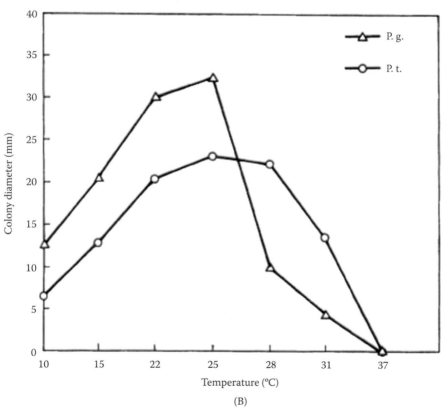

(B)

7.26   MORPHOLOGIES: *PYRENOCHAETA GENTIANICOLA*

A–B. Pycnidia formed on gentian straw in water agar used as natural medium. Note several aerially-developed necks and embedded pycnidia on the tissue surface.

C. Pycnidia formed in agar medium.

D. Crushed pycnidium. Note the neck and the setae around.

E. Conidiophores and conidia.

F. Conidia.

G. Close-up of conidiophores.

(*P. gentianicola* colony dark yellowish green. Pycnidia necked with setae. Conidiophores ampliform. Conidia elliptical, 4 × 1.1 µm. Sclerotia or resting cells: catenulate, rough.)

(Watanabe and Imamura, 1995)

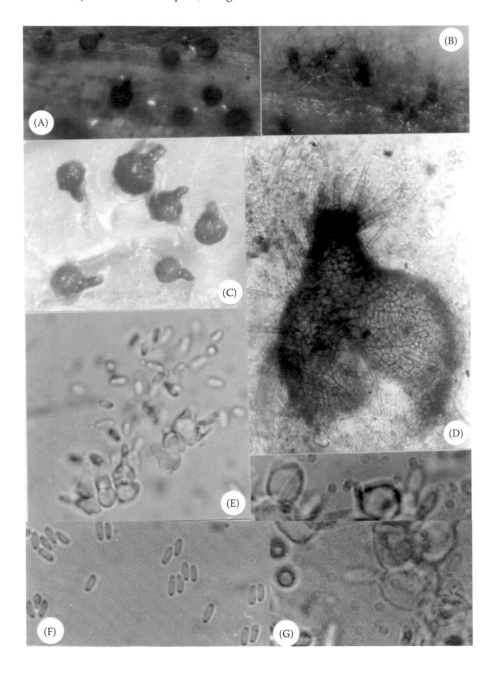

7.27   MORPHOLOGIES: *PYRENOCHAETA TERRESTRIS*

A. Crushed pycnidium.

B. Close-up of Figure **A**.

C. Conidiophores.

D. Conidia.

(*P. terrestris* colony gray, reverse brown to black. Pycnidia solitary or aggregated, globose or subglobose, dark brown to black, 123.5–296.5 μm in diameter, non-necked, with setae around, ostiolate. Setae brown, triple-septate, 50–142.5 μm long, 2–4.5 μm wide at base, up to 2 μm wide at apex. Conidiophores obpliform, 4.6–6.6 × 1.3–2.8 μm. Conidia hyaline, elliptical, 4.8–5.5 × 2.3–3 μm. Sclerotium-like structures 25–85 μm in diameter)

(Watanabe and Imamura, 1995)

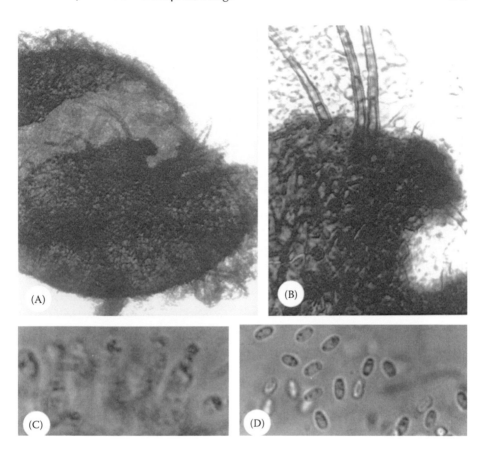

7.28   MORPHOLOGIES: *PYRENOCHAETA GLOBOSA*

A. Papillate pycnidium.

B–F, H. Pycnidia with single (**B–F**) or two ostioles (**C, H**).

G. Conidiophores, conidia, and a part of the pycnidial wall.

I–J. Setae around the ostiole and conidia.

K. Conidia.

(The fungus [isolate 84–523 (ATCC201314)] from Japanese black pine seeds was characterized by ostiolate pycnidia with thick-walled setae, and phialosporous globose conidia.)

(Watanabe, 1992)

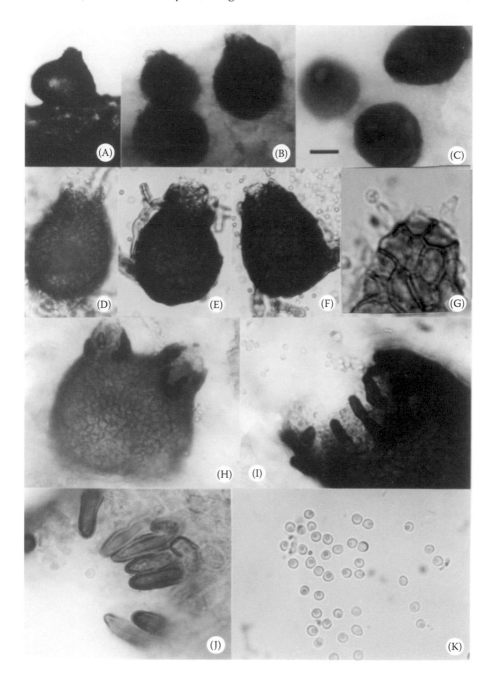

7.29 POTATO ROOT ROT PATHOGEN, *RHIZOCTONIA SOLANI*

Rhizoctonia species are common as one of the soilborne pathogens together with *Fusarium* and *Pythium*. Known as the damping-off pathogen, they also cause root damages of various plants. This fungus is also seedborne and contaminated or infected with various commercial seeds.

7.30 DISEASES: POTATO ROOT ROT PATHOGEN AND SPECIMEN

A. Healthy (left) and damaged potato with black blemish (right) caused by *R. solani*.

B. Underground damage of seed potato with sclerotia of *R. solani* (arrow) adhering to the lateral roots.

C. Dried specimen of *R. solani* (isolate 91–33).

(Watanabe, 1992)

(A)

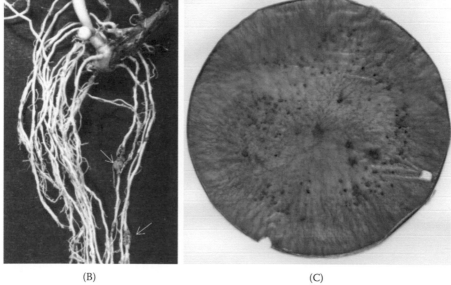

(B)   (C)

7.31   MORPHOLOGIES: *RHIZOCTONIA SOLANI,* DAMPING-OFF PATHOGENS IN NURSERY

A. *R. solani* colony from commercial kidney bean seed.

B. *Rhizoctonia* isolates from strawberry roots.

C. *Rhizoctonia* isolates from sugar cane roots.

D–E. Various *Rhizoctonia* colonies isolated from flowering cherry seeds.

(Watanabe, 1992)

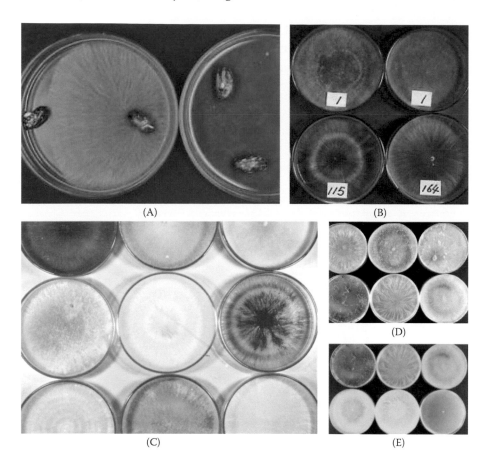

(A)

(B)

(C)

(D)

(E)

7.32 MORPHOLOGIES AND DISEASES: *RHIZOCTONIA SOLANI* AND RELATED SP., PATHOGENICITY

A. Pathogenicity of *R. solani* from commercial kidney bean seeds to bean hypocotyls.

B. Collapsed cucumber seedlings in the infested soil with *R. solani* (left) and healthy seedlings in noninfested soil (right).

C–E. Hyphae of *R. solani* often with monilioid cells (**D**) (isolates 83–417 (**C, D**) and 85–65 (**E**)).

F. Various colonies of *Rhizoctonia* isolates from strawberry roots in Tsukuba.

Top row: *R. fragariae* (isolate 98–42) and *R. solani* (isolate 98–43).

Bottom row: *R. solani* (isolate 98–44) and *Rhizoctonia* sp. (isolate 98–112).

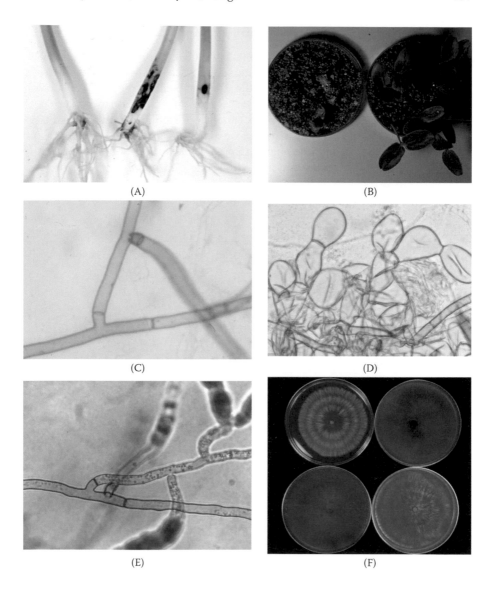

(A)

(B)

(C)

(D)

(E)

(F)

7.33   DISEASES: CHINESE CABBAGE YELLOWS

Chinese cabbage yellows have been serious since 1966 in the fields in Nagano, Japan. Some 10 ha of the fields were damaged due to the disease. The symptoms become serious near mature stage. The leaves of the diseased plants turned yellow in color, and the plants were a little stunted. The central heads were never packed even near harvest time in these plants. Their tap roots and stems became dark brown or blackened in the vascular tissues. The pathogen was initially suspected to be *Fusarium*. The most frequently associated fungus, *Cephalosporium* sp. (= *Acremonium* sp.) (65% of the total 143 isolates) was not pathogenic. *Verticillium albo-atrum* was also frequently associated with the damaged vascular bundles (26% of the isolates). It showed pathogenicity onto three varieties of Chinese cabbages tested. The discoloration of the root and stem generally observed in some 40 days after seeding in the artificially infested soil at 20°C.

A. Chinese cabbage yellows found at the fields, in Nagano, Japan. Note blight yellowed stunted cabbages.

B. A diseased cabbage naturally infected with *Verticillium albo-atrum* showing vascular discoloration of the root and stem tissue in longitudinal section.

C–D. Hyphae observed in the vascular tissues in cross (**C**) and longitudinal sections (**D**).

(Watanabe et al., 1973)

(A)   (B)

(C)   (D)

7.34   DISEASES: CHINESE CABBAGE YELLOWS CAUSED BY *VERTICILLIUM ALBO-ATRUM*

A–B. Pathogenicity of *V. albo-atrum* to two varieties (**A**: Chirimen, **B**: Yokozuna) of inoculated Chinese cabbages (central and right) in relation to uninoculated control plants (leftmost).

C–D. Darkening of the vascular bundles (cross section (**C**) and longitudinal section (**D**)) of Chinese cabbage roots grown in the infested and noninfested soil.

E–F. Pathogenicity of *V. albo-atrum* to inoculated potted cabbage and tomato seedlings (**E** and **F**, right) in relation to uninoculated healthy seedlings (**E** and **F**, left). Note stunted, vein-twisted damaged or dried-up plants.

Two isolates of *V. albo-atrum* were pathogenic to three varieties of Chinese cabbage in three different inoculation methods (inocula poured near each seedling, placed near seeds, or mixed soil). For example, 17% of 123 plants of the Chirimen variety were blackened in the vascular tissues with the isolate V–1, and 30% of them were infected.

(A)

(B)

(C)

(D)

(E)

(F)

7.35   DISEASES: BELL PEPPER STUNTED BY *VERTICILLIUM ALBO-ATRUM* INOCULATION

A. Healthy and stunted bell pepper seedlings that are noninfested (left) and infested with *Verticillium albo-atrum* (isolate 73–314) (center and right).

(A)

7.36 MORPHOLOGIES: CHINESE CABBAGE YELLOWS CAUSED BY *VERTICILLIUM ALBO-ATRUM*

A. Ten-day-old PDA colony of *V. albo-atrum* (isolate V–1).

B. Conidiophore bearing verticillate branches and terminal spore masses.

C. Conidia.

D. Dark resting mycelium formed in ten-day-old PD broth culture.

Twenty percent of 20 plants, and 22% of 18 plants became blackened in the vascular tissues in the inoculum quantity of 0.2, or 0.4 g per kg of soil, respectively. However, no vascular bundles discolored in any test at 0.1 g, or control. Infection rates at the inoculum levels of 0.1, 0.2, or 0.4 g were 9.1, 5.0, or 27.8%, respectively.

Disease incidence was measured in relation to inoculum quantity (Figure A, p. 191) and progression of time (Figure B, p. 191).

An average of 12.7% of 102 plants tested were infected in ten days after seeding at 20°C, and 31.3 or 75.0% of the plants were infected, in 20 or 30 days, respectively (Figure B, p. 191).

(*V. albo-atrum* colony dark gray. Conidiophore bearing hyaline verticillate branches and terminal spore masses. Conidia elliptical, hyaline, 5.3–8.8 × 2.5–4.5 μm. Resting cells thick-walled, 8.8–17.5 × 7.5–13.8 μm)

(Watanabe et al., 1973)

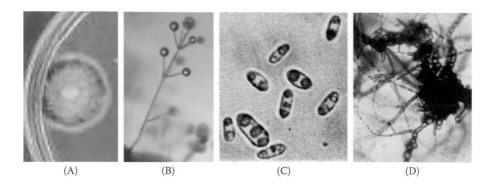

(A)   (B)   (C)   (D)

7.37 MORPHOLOGIES: CHINESE CABBAGE YELLOWS CAUSED BY *VERTICILLIUM ALBO-ATRUM* (CONTINUED)

A. Rates of vascular discoloration (grey bar), and infection (white bar) of Chinese cabbage in relation to inoculum quantity in the artificially infested soils.

B. Infection rates of Chinese cabbage with *Verticillium abo-atrum* in relation to time after seeding in the artificially infested soils.

(Watanabe et al., 1973)

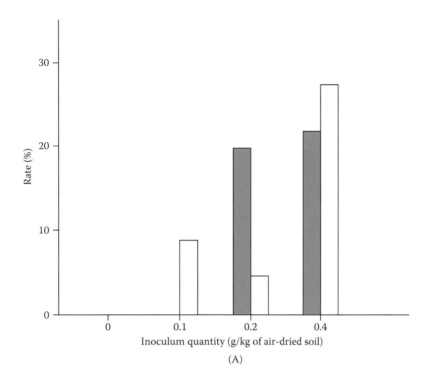

(A)

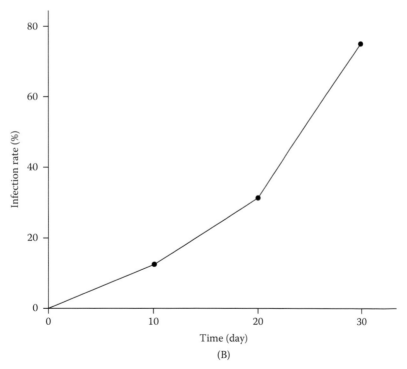

(B)

7.38 DISEASES: ZONATE LEAF SPOT OF SORGHUM

Gloeocercospora sorghi (isolate 74–492 (MAFF425368)) was recovered from 68% of sorghum seeds collected from the severely-diseased fields in Ebina, Kanagawa. The emergence rate of these seeds grown in autoclaved soil was as low as 20%, and 80% of the emerged seedlings were diseased with lesions on their leaves.

The conidia were variable in size, 15.0–245.0 × 2.5–4.5 μm with 1–24 septates forming, germinated 100% within 4 hours on water agar at 20–30°C. Sclerotia prepared from 30-day-old culture (spherical, black, hard, 98.8–230.0 × 74.1–172.9 μm) also germinated 96% of the time, and 22% of the sclerotia formed sporodochia directly on them on WA four days after seeding. Experimental results suggest that the fungus is transmitted through infected plants.

A. Lesions on sorghum leaves.

B. Lesions formed around inocula of *G. sorghi* on sorghum leaves.

(Watanabe and Hashimoto, 1978)

(A)   (B)

7.39 MORPHOLOGIES: ZONATE LEAF SPOT PATHOGEN OF SORGHUM, *GLOEOCERCOSPORA SORGHI*

A. Spore masses and sclerotia formed on the sorghum straw.

B. Spore mass.

C. Conidiophores and conidia in culture.

D–E. Germination of conidia on water agar two hours (**D**) and 24 hours after treatment (**E**).

F. Sclerotium germination forming sporodochium.

G. Macro- and microconidia.

H. A part of conidium.

I. Sclerotia.

(Watanabe and Hashimoto, 1978)

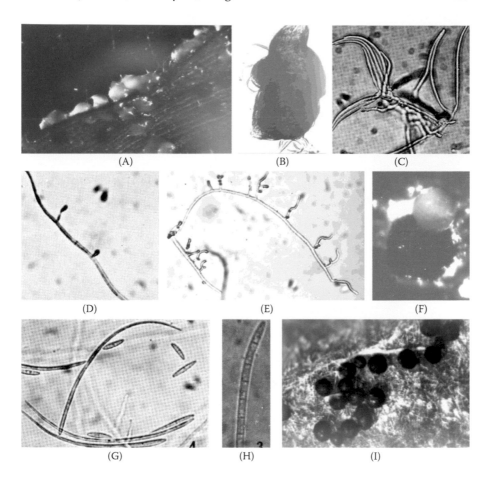

(A)   (B)   (C)

(D)   (E)   (F)

(G)   (H)   (I)

8 Various Related Topics on Soil Fungi

8.1   POOR RATOONING OF SUGARCANE IN TAIWAN: OUTLINE

The poor emergence or poor crop of ratoon cane in relation to the yield decline had been problematic in Taiwan around the 1970s. Sugarcane cultivation there was started or based on the canes left after harvest and cultures were succeeded continuously. The emergence rates were very often less than 50%. In the severely damaged fields, the nonemerged or emerged ratoon canes had generally poor root systems. To clarify the poor crops or yield losses, damaged canes were checked in detail scientifically, including plant nutrition, crop science, and plant protection.

Damaged parts were dissected and tissues were assayed for the microorganisms. Various fungi were detected using more than 34 canes of 12 fields, and 1610 tissue pieces from roots, buds, or shoots and seed pieces of test plants. The fungal flora associated with the underground parts of sugarcanes were clarified. Among a total of 1010 isolates belonging to 44 genera, fungi selected were tested for their pathogenicity. Each test fungus was grown in PD broth for ten days and they were used as inocula after removing the culture residue. Infested soils were prepared by burying them in autoclaved soil. Healthy cut canes (single eye cuttings, 13 cm long) were buried in soil longitudinally and cultured under greenhouse conditions at 30°C for a month. Pathogenicity was evaluated on the basis of emergence rate and top height of test plants. By testing pathogenicity of 49 isolates, *Thielaviopsis paradoxa* (isolate 72-X49-378 (ATCC32928)) and *Pythium catnulatum* (isolate 72-X99 (ATCC38892)) were the most pathogenic, and they were believed to be the causes of the diseases.

T. paradoxa conidiophores contain hyaline, simple, cylindrical-bearing conidia and chlamydospores. Conidia are phialosporous, single-celled, cylindrical, or spindle-shaped. The cylindrical conidia are hyaline to pale brown and thin-walled, while the spindle-shaped conidia are brown and thick walled. Chlamydospores are dark brown, ellipsoidal, thick-walled, and granulate. Conidiophores are 121.6–350.3 μm long. Conidia: cylindrical 8.5–16.3 × 2.5–4.7 μm; spindle-shaped 9.2–12.5 × 5.1–7.5 μm. Chlamydospores are 13.3–20.5 × 9.7–11.7 μm.

P. catnulatum forms lobate sporangia, aggregates of globose cells or conidia, and conidia germinated directly. Oogonia bear 1–6 crook-necked antheridia per oogonium, mono- or diclinously. Oospores are plerotic, with thick oospore walls. Sporangia are 17–20 μm wide, and vesicles are approximately 55 μm in diameter. Oogonia are 26.4–42.5 μm in diameter. Oospores are 18.7–35 μm in diameter; the oospore wall is 1.2–3.8 μm wide. Antheridia are 6.2–6.5 μm wide.

(Watanabe et al., 1974)

8.2   POOR RATOONING OF SUGARCANE IN TAIWAN: DISEASED FIELD AND INOCULATION

A. Sugarcane field with poor ratooning in Taiwan.

B. Poor emergence of ratoon cane in the field.

C. Damaged seedlings grown in the autoclaved potted soil artificially infested with *Thielaviopsis paradoxa* (right two pots), and healthy ones (left two) grown in the autoclaved soil, 35 days after planting under the growth chamber conditions.

D. Underground parts of damaged (right three plants) and healthy canes (left three plants) removed from potted soil.

E. Damaged seedlings grown in the autoclaved potted soil artificially infested with *Pythium catenulatum* (right two pots), and healthy ones (left two) grown in the autoclaved soil, 35 days after planting under the growth chamber conditions.

F. Underground parts of damaged (right three plants) and healthy canes (left three plants) removed from potted soil.

(Watanabe et al., 1974)

(A)

(B)

(C)

(D)

(E)

(F)

8.3   POOR RATOONING OF SUGARCANE IN TAIWAN: MORPHOLOGIES OF THE CAUSAL AGENTS

1. *Pythium catenulatum* (**A–F**) (isolate 72-X99 (ATCC38892))

    A. Lobate sporangium.

    B. Catenulate globose cells.

    C. Direct germination of sporangium.

    D. Encysted zoospores.

    E–F. Oogonia, antheridia, and oospores.

2. *Thielaviopsis paradoxa* (**G–H**) (isolate 72-X49-378 (ATCC32928))

    G. Conidia and chlamydospores.

    H. Conidia and a part of phialide.

3. Grass cicada (*Mogannia hebes*, a large flylike insect with transparent wings) found in cane fields everywhere in Taiwan.

    I–J. Cicadas related to the poor ratooning of sugarcane.

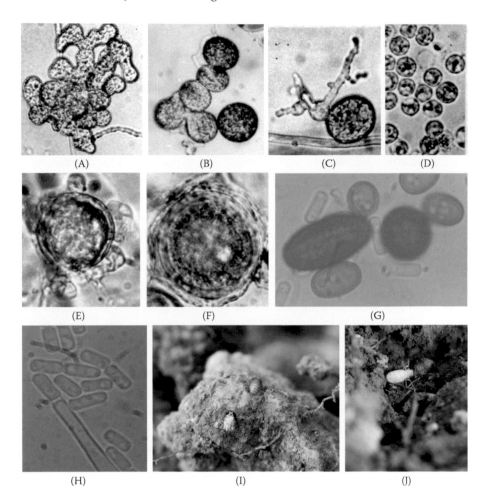

(A) (B) (C) (D)

(E) (F) (G)

(H) (I) (J)

8.4 DISEASES: HEALTHY STRAWBERRY SEEDLINGS RAISED IN PATHOGEN-FREE SOIL

Strawberry plants grown in the natural soil were often stunted and root-rotted. In the sterilized soil, plants grew well. This fact has been known in Canada and other countries since the 1930s.

Pathogen-free soil was prepared by autoclaving soil at 100°C for four hours, and stocked in a wooden box as pasteurized soil. Fungi were not recoverable soon after autoclaving, but after 21 days, *Trichoderma, Cunninghamella, Monilia,* and *Penicillium* were isolated at the rates of 7.8%–37.3%, but the former three genera were replaced gradually by *Penicillium.*

The seedlings grown in the potted pasteurized soil set in the field were larger than those in the potted natural soil in size, heavier in fresh weight, and more numerous in the root number, and less in the diseased root rates. The grand average weight per plant was 15.5 g in the former soil versus 8.7 g in the latter, and the healthy root rate was 46.8% versus 28.3%.

Penicillium spp. were predominant among 28 fungus genera recovered from the strawberry roots grown in the potted pasteurized soil set in the field, whereas *Rhizoctonia* and *Fusarium* spp. were dominant among 30 genera from the roots in the potted natural soil (**B**). The plants increased in weight in the pasteurized soil compared with those in the natural soil 132 to 137 days after transplanting (**A**).

The reduction of the root damage may be attributable to the potential soilborne pathogens including *R. solani* by the antagonism of *Penicillium* spp.

A. Growth of strawberry seedlings under natural environmental conditions 132 days after unrooted runners were struck in the potted soil in the field. Note the two clay pots in the soil.

B. Fungi associated with strawberry roots grown in the potted pasteurized soil (blank bars), and natural soil (dark bars), and their isolation frequencies. The parenthesized figures (1, 2, 3) indicate the number of times the respective fungi were recovered, out of three total experiments.

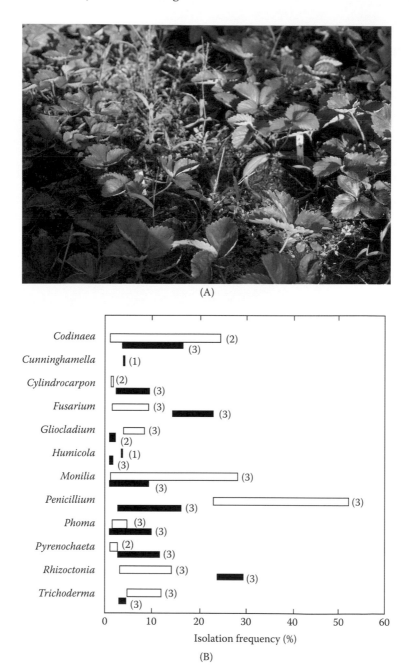

(A)

(B)

Isolation frequency (%)

8.5   MORPHOLOGIES: HEALTHY STRAWBERRY SEEDLINGS RAISED IN PATHOGEN-FREE SOIL

A. Strawberry roots grown in the potted pasteurized (left) and natural soil (right) 122 days after treatment.

B. Antagonism of *Penicillium* spp. (isolates 78–2601, center; and 78–2610, right) to *Pythium splendens* (lower row) and *Rhizoctonia solani* (top row) in relation to single control cultures of both fungi (left) 48 hours after inoculation.

C–E. Antagonism shown by disintegrated hypha (**C**) and apical inflation (**D**) of *P. splendens,* and intermingled hyphae (**E**) of *R. solani* near the inhibition zone of dual cultures.

(Watanabe and Inoue, 1980)

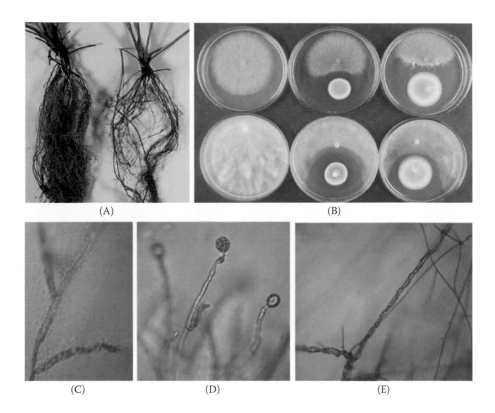

(A)

(B)

(C)

(D)

(E)

8.6   MORPHOLOGIES: NEMATODE ENDOPARASITE, *VERTICILLIUM BALANOIDES*

Verticillium balanoides was isolated from nematodes associated with collapsing Japanese red pine trees in Tsukuba, Japan.

Verticillium balanoides (isolate 80–33 (MAFF237890)).

A–B. Sporulation on dead nematodes.

C. Verticillate phialides and conidia. Note spore masses terminally on simple, paired, or verticillate phialides and maize-kernel-like or acorn-shaped conidia.

(Watanabe, 2000)

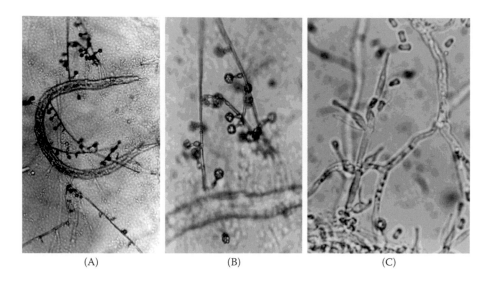

(A)   (B)   (C)

8.7   MORPHOLOGIES: NEMATODE ENDOPARASITE, *VERTICILLIUM SPHAEROSPORUM* VAR. *BISPORA*

A–B. Nematode parasitized by *V. sphaerosporum* (isolate 76–500).

B: Close up of **A**.

C. Conidia attached around the nematode buccal region.

D–E. Hyphae inside disintegrated nematode (**D**) and sporulation (**E**).

F. Simple conidiophore bearing cylindrical conidia and spore mass composed of globose conidia.

G. Phialide and globose conidium.

H. Verticillate conidiophore bearing terminal spore masses.

I. Cylindrical and globose conidia.

(Watanabe, 1980)

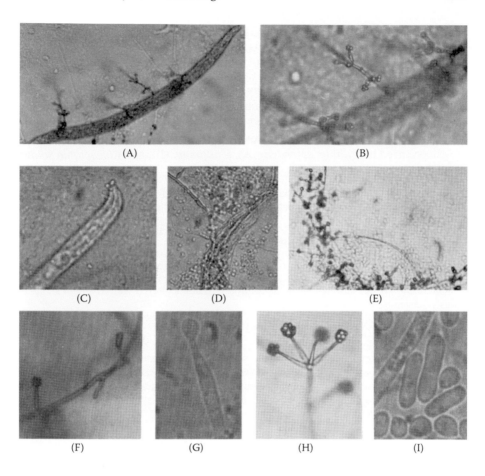

(A)

(B)

(C)

(D)

(E)

(F)

(G)

(H)

(I)

8.8　MORPHOLOGIES: NEMATODE ENDOPARASITE, *VERTICILLIUM SPHAEROSPORUM* VAR. *BISPORA* (CONTINUED)

A new variety of *V. sphaerosporum* (isolate 76–500) isolated from conidia on dead nematodes on water agar originally plated with decaying strawberry root tissues, showed endoparasitic effects against some nematodes (*Aphelenchoides* sp., *Cephalobus* sp., and *Panagrolaimus* sp.) tested. It also showed antagonism to various fungi from the strawberry root rhizosphere. More than 50% inhibition of radial growth occurred in *Pythium splendens* (**A**), *P. ultimum* (**B**), and *Rhizoctonia solani* (**C**) and five other fungi tested.

A–B. Antagonism of *V. sphaerosporum* on *Pythium splendens* (isolate 73–192 (ATCC36444)) (**A**) and *P. ultimum* (isolate 74–901 (ATCC36443)) (**B**).

V. sphaerosporum was cultured singly as control (top left), or cocultured with each test fungus simultaneously (top right), three days later (lower left), or five days later (lower right).

C. Antagonism of *V. sphaerosporum* on *Rhizoctonia solani* (isolate 78–2101).

V. sphaerosporum was cultured singly as control (left) or cocultured with *R. solani* simultaneously (center), or three days later (right).

(Watanabe, 1980)

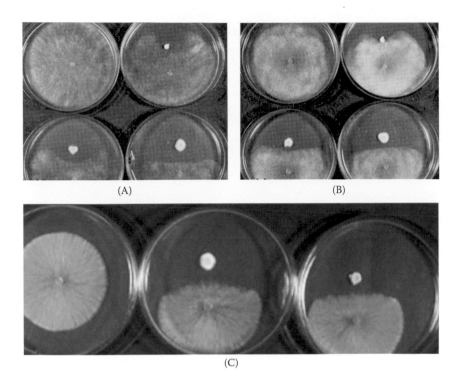

(A)

(B)

(C)

8.9   MORPHOLOGIES: *TAENIOLELLA PHIALOSPERMA* AS A PROMISING BIOCONTROL AGENT

Taeniolella phialosperma

A. Twenty-eight-day-old PDA colonies of two isolates (left, isolate 70–1070; right, isolate 73–466).

B. *Taeniolella* state.

C. *Taeniolella* and *Phialophora* states on the same hypha.

D. *Phialophora* state developed from phragmospore in *Taeniolella* state.

E. *Phialophora* state.

F. Branched and catenulate conidia in *Taeniolella* state.

G. Catenulate conidia in *Taeniolella* state.

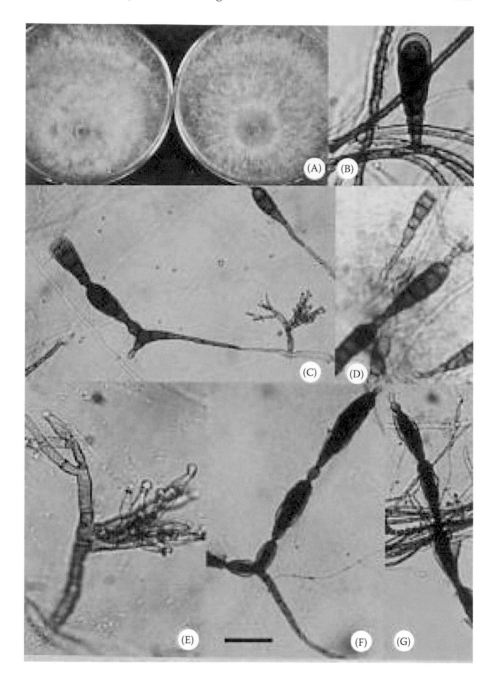

8.10   DISEASES: *TAENIOLELLA PHIALOSPERMA* AS A PROMISING BIOCONTROL AGENT

During an etiological study of strawberry stunt disease in Japan, an interesting Hyphomycetous fungus was isolated from the rhizosphere soil. The similar fungus was also isolated from the paddy field soil in Hachijō-jima, Tokyo, Japan. This fungus was unique and characterized by a distinctive *Taeniolella* state with *Phialophora* synanamorph. It was named *T. phialosperma*. Two isolates from the strawberry rhizosphere isolate (TW 73–466) and paddy field soil isolate (70–1070) are described.

A. Black pine seedlings grown in noninfested soil (left five dishes), infested soil with *Dematophora necatrix* (isolate 84–373), and infested soil treated with *T. phialosperma* isolates (third column, isolate 70–1070; far right, isolate 73–466).

B. Healthy cucumber seedlings grown in noninfested soil (control) (left two dishes), collapsed seedlings in infested soil with *Pythium aphanidermatum* (central two dishes, isolate 82–992), and healthy seedlings in the infested soil treated with *Taeniolella phialosperma* (far right, isolate 73–466).

(Watanabe, 1992; Watanabe et al., 1977)

(A)

(B)

8.11 DISEASES: *TAENIOLELLA PHIALOSPERMA* AS A PROMISING BIOCONTROL AGENT (EXPERIMENTAL DATA)

TABLE 8.1
Control of White Rot (Pathogen: *Dematophora necatrix*) of Black Pine Seedlings in Natural Soil and Autoclaved Soil by Seed Treatment with Antagonistic *Taeniolella phialosperma* Isolates (70–1070, 73–466)

Experiment	Damaged no./Emerged no.		Emergence %	Healthy Seedlings %
	1	**2**	**Av.**	**Av.**
Autoclaved soil:				
Control	0/46	0/46	92	92
Path.	23/48	32/43	91	36
Path. + A70–1070	1/44	9/46	90	80
Path. + A73–466	0/42	0/47	89	89
Natural soil:				
Control	0/46	2/46	92	90
Path.	5/49	6/49	98	87
Path. + A70–1070	1/49	4/47	96	91
Path. + A73–466	6/46	2/50	96	88

Note: Results were obtained 30–40 days after sowing.

8.12 *SORDARIA* SPP., AND SOME SEED-ASSOCIATED FUNGI AS PROMISING BIOCONTROL AGENTS

The coculture method has often been used for screening the antagonists among numerous candidates against any particular pathogens. The inhibition zones were usually formed between both fungi. For example, *Pythium aphanidermatum* and *Dematophora necatrix* were reduced for their growth in coculture with *Sordaria* spp. by 35%–60% and 22.2%–33.5%, respectively, compared with their single control cultures.

During the study of rhizomorph production of *Armillaria mellea* by the influence of various fungi, inhibition zones were formed between *A. mellea* and *Sordaria fimicola* in coculture on PDA, together with perithecia around the margin of inhibition zones.

The effects of the antagonists were evaluated based on the emergence, and healthy seedling rates in the soil over cultures of pathogens and antagonists in various combinations in plates, or in the infested soils, directly applied by pouring PD broth cultures of antagonists or sowing the seeds treated with antagonists. Biocontrols of the soilborne plant diseases by *Sordaria* spp. were almost successful in any treatment.

Sordaria spp. and *A. mellea* were confirmed for their mutual influence on the stimulation of the rhizomorph of *A. mellea,* and perithecia and ascospore formation of *Sordaria* spp. in coculture. The stimulation of perithecium and ascospore production may be related to physiologically active substances. The vitamin effect was confirmed in subsequent work.

(Watanabe, 1986, 1989, 1990, 1991, 1997)

8.13   *SORDARIA* SPP., AND THEIR EVALUATION AS BIOCONTROL AGENTS AGAINST SOILBORNE PLANT DISEASES

A–B. *Sordaria fimicola* (isolate 85–84): Perithecia (**A**), ascus and ascospores (**B**).

C. *S. tamaensis* (isolate 85–72), Crushed perithecium with discharged asci and ascospores.

D. Antagonism of *Sordaria* spp. against *Dematophora necatrix* by coculture method, showing *D. necatrix* in single culture as the control, in coculture with *S. nodulifera*, *S. fimicola* (three isolates), or *S. tamaensis* from top left to lower right, seven days after inoculation.

E. Komatsuna *(Brassica campestris)* seedlings grown in the *P. aphanidermatum*-infested soil (center) were collapsed, but partly survived after growing the *S. tamaensis* (antagonist)-treated seed (right, isolate 85–89), and were almost healthy in noninfested soil (left).

F. The cucumber seedlings grown in the *P. aphanidermatum*-infested soil (center) were completely collapsed, but partly survived in the *S. nodulifera* (antagonist)-treated soil (right), and were almost healthy in noninfested soil (left).

G. The Japanese black pine seedlings grown in the *D. necatrix*-infested soil. In the nontreated seeds, seedlings (center) were poor in growth, but in the *S. nodulifera*-treated seeds (right), and in nontreated seeds (left) were almost healthy.

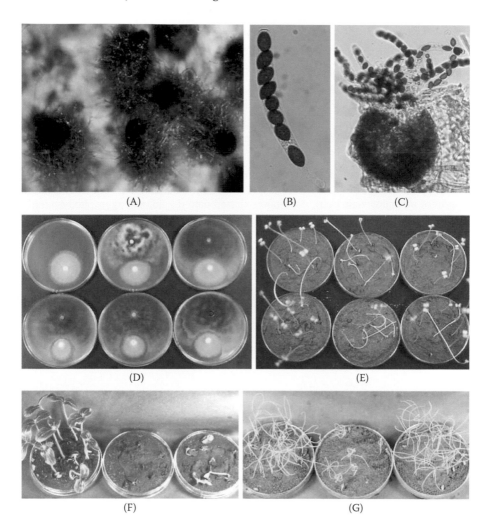

8.14   TRIALS OF *SORDARIA* spp. AS PROMISING BIOCONTROL AGENTS AGAINST *ARMILLARIA MELLEA*

A–B. *Armillaria mellea* (isolate 85–153 (MAFF45285)) formed rhizomorphs on 9 mL PDA plate supplemented with 1 mL of heat-treated (**A**, left), noncultured PD broth (**A**, right) and nonheated PD culture broth of *Macrophoma* sp. (isolate 84–25) (**B**).

C–D. *A. mellea* formed rhizomorphs on PDA supplemented with 1 mL of heat-treated (left) and nonheated PD culture broth of *Sordaria* sp. (right). Note the inhibition zone formed around *A. mellea*. Dried specimen (**D**).

E. Inhibition zone formed around *A. mellea* inoculum on PDA supplemented with 1 mL of nonheated PD culture broth of *Sordaria* sp. Note perithecia and ascospores formed in the margin of inhibition zone (left). Ascospores ejected onto the lid, and aggregated centrally (right). Pictures were taken ten days after treatment.

F. Ascospores ejected.

(Watanabe, 1986, 1991, 1997)

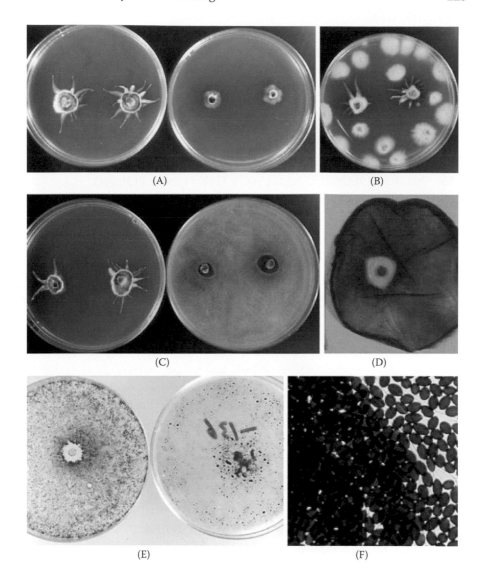

(A)

(B)

(C)

(D)

(E)

(F)

8.15 MORPHOLOGIES: *SORDARIA FIMICOLA* INFLUENCED FOR ITS PERITHECIUM AND ASCOSPORE PRODUCTION BY *ARMILLARIA* SPP. AND VARIOUS FUNGAL SPECIES

Armillaria mellea and *A. tabescens* stimulated the production of the perithecia and ascospore production of *Sordaria fimicola* over-agar-culture, and culture-broth assay method shown by the coculture, spore-plate, over-agar-culture, and culture broth assay methods (Watanabe, 1997).

8.16 MORPHOLOGIES: NOTEWORTHY FUNGI FOUND DURING STUDIES ON SOILBORNE DISEASES

Among mycological works during the studies on soilborne diseases, various new fungi were detected, named as new fungi, and published. Some representative fungi were tabulated alphabetically in Table 8.2 (p. 232).

As etiological studies, some diseases are associated with these new fungi. Some of them include *Nodulisporium melonis* associated with melon root rot, *Pyrenochaeta gentianicola* associated with gentian pink root rots.

Some fungi were first reported in Japan, including *Cylindrocladium colhounii*, *C. floridanum* associated with *Phellodendron amurense*'s decline, *Vericillium albo-atrum* with Chinese cabbage yellows, and various nursery damping-off diseases caused by *Pythium deliense*, *P. splendens*, *P. myriotylum,* and *Phytophthora megasperma*.

For inoculation tests to verify their pathogenicity, clean healthy seeds were essential for this test. Thus fungi associated with these seeds were also surveyed. As a result, some were found new. They were *Pyrenochaeta globosa* associated with black pine seeds, *Sordaria nodulifera* and *S. tamaensis* associated with flowering cherry seeds. Several new fungi were found to be associated with damaged plants.

8.17 MORPHOLOGIES: *COPRINOPSIS CINEREA* FROM RICE HUSKS FORMING SCLEROTIA

Sclerotia were formed in agar culture by a fungus with clamp connections isolated from rice husks in Tsukuba, Japan. The sclerotia were brown, globose to ellipsoidal, small, up to 200 μm in diameter, and composed of external rind tissue and internal medulla tissue. Such tiny sclerotia have not been commonly reported among basidiomycetous fungi in literature. The fungus was identified as *Coprinopsis cinerea* on the basis of morphological characteristics together with molecular analyses. Three reference strains of *C. cinerea* formed sclerotia similarly under identical conditions.

Sclerotium morphology of *Coprnopsis cinerea* (isolate 06–150).

A. Rice husks collected in Tsukuba.

B. A fungus formed clamp connections in agar culture.

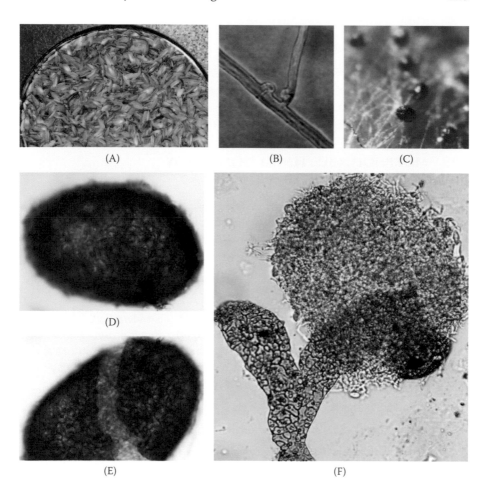

(A)

(B)

(C)

(D)

(E)

(F)

(previous page)

C. Globose sclerotia formed on water agar.

D, E. Crushed ellipsoidal sclerotium, composed of rind and medulla tissue.

F. Sclerotium composed of pseudoparenchymatous rind (lower part) and prosenchymatous medulla tissue (upper part).

(Watanabe et al., 2011)

8.18   MORPHOLOGIES: FUNGI ISOLATED BY TW AND REGISTERED AS NEW FUNGI

After harvest, rice husks left in the rice fields have been troublesome on the basis of plant protection aspects, because they must be the source of plant pathogens and animal damages.

Several noteworthy fungi were found among these rice-associated fungi. They were eight ascomycetous fungi including three new species: *Didymosphaeria* sp., *Nectriopsis* sp., and *Setosphaeria* sp.

Coprinopsis cinerea and another basidiomycete were also identified. There were 22 kinds of anamorphic fungi including *Phleospora graminiarum*, *Phoma* sp., *Pyrenochaeta* sp., and *Stachybotrys bisbi*. Rice husk fungus floras were rich and diverse. Although *G. graminis* var. *graminis* has been known as a take-all pathogen of cereals, the fungus (isolate 06–116) was not pathogenic to rice seedlings.

(Watanabe and Nakamura, 2008)

A. *C. cinerea* (isolate 06–150) sporulated in around two-month-old PDA culture.

B. Elongated stalk and umbrella opened with black spherical spore print nearby.

C. Basidiospores.

(Watanabe et al., 2011)

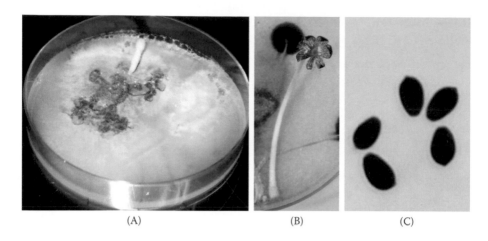

(A) (B) (C)

8.19   MORPHOLOGIES: FUNGI ISOLATED BY TW AND REGISTERED AS NEW FUNGI (CONTINUED)

1. *Didymosphaeria* sp. (isolate 06–134)

    A. PDA colony.

    B. Ascocarp.

    C. Asci and ascospores.

    D. Ascospores.

2. *Nectriopsis* sp. (isolate 06–148)

    E. PDA colony.

    F. Ascocarps.

    G. Asci and ascospores.

    H. Ascospores.

3. *Setosphaeria* sp. (isolate 06–146)

    I. PDA colony.

    J–K. Ascocarps.

    L–M. Asci and ascospores.

4. *Gaeumannomyces graminis* var. *graminis* (isolate 06–116)

    N. PDA colony.

    O. Ascocarp.

    P. Hyphopodium.

    Q. Asci and ascospores.

    R. Ascospores.

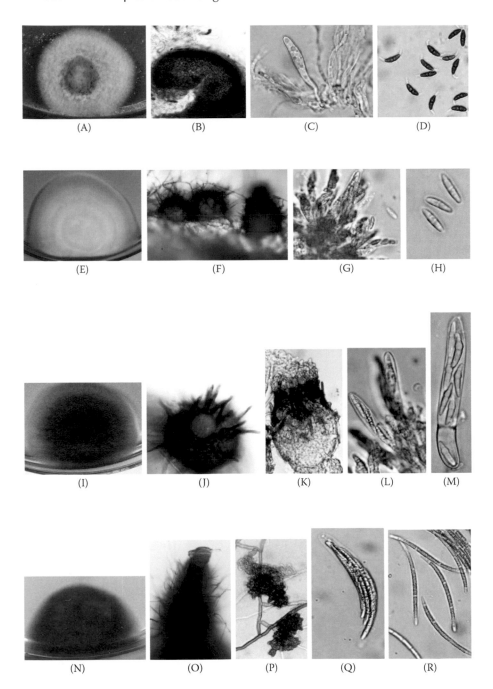

(A)   (B)   (C)   (D)

(E)   (F)   (G)   (H)

(I)   (J)   (K)   (L)   (M)

(N)   (O)   (P)   (Q)   (R)

8.20 MORPHOLOGIES: FUNGI ISOLATED BY TW AND REGISTERED AS NEW FUNGI (CONTINUED)

5. *Mycosphaerella ligulicola* (isolate 06–122)

    A. PDA colony.

    B–C. Ascocarps.

    D. Asci and ascospores.

    E. Conidiocarp.

    F. Conidia.

6. *Phaeosphaeria nigrans* (isolate 06–144)

    G. Ascocarps.

    H. Asci and ascospores.

    I. Ascospores.

7. *Phaeosphaeria oryzae* (isolate 06–145)

    J. Ascocarps.

    K. Ascocarp wall.

    L. Asci and ascospores.

    M. Ascospores.

8. *Sphaerullina oryzae* (isolate 06–112)

    N. PDA colony.

    O. Ascocarps.

    P–Q. Asci and ascospores.

    R. Ascospores.

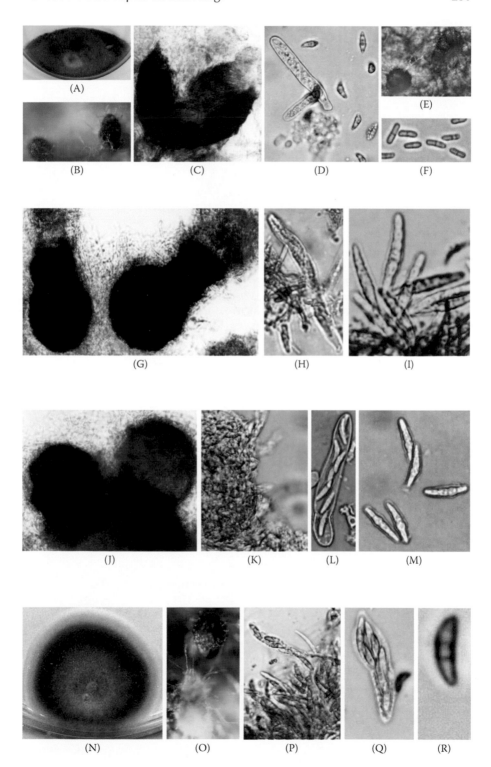

(A)

(B)   (C)   (D)   (E)

(F)

(G)   (H)   (I)

(J)   (K)   (L)   (M)

(N)   (O)   (P)   (Q)   (R)

TABLE 8.2
Fungi Isolated by TW and Registered as New Fungi

New Fungus	TW	Holotype	Reference
Acremonium macroclavatum	00–50	(MAFF238162, CBS123922)	*Mycoscience* 42: 591–593, 2001
Cylindrocarpon boninense	00–62	(MAFF238163)	*Mycoscience* 42: 593–594, 2001
Cylindrocladium tenue	92–246	(ATCC201311, MAFF425366, NBRC32533)	*Mycologia* 86: 155, 1994
Dactylella chichisimensis	00–315	(MAFF238165)	*Mycoscience* 42: 633–635, 2001
Hyphodiscosia radicicola	77–94	(ATCC200611, MAFF425311, NBRC32540)	*Mycologia* 84: 113–114, 1992
Irpicomyces cornicola	93–107	(ATCC200996, MAFF425383)	*Mycologia* 35: 105–108, 1994
Mortierella tsukubaensis	98–120	(ATCC204319, MAFF237778)	*Mycol. Res.* 105: 506–550, 2001
Mucor hachijoensis	70–1179	(ATCC201000, MAFF425203)	*Mycologia* 86: 692–695,1994
Mucor meguroense	92–268	(ATCC200999, MAFF425385)	*Mycologia* 86: 692–695,1994
Myrothecium dimorphum	01–250	(MAFF238296)	*Mycoscience* 44: 284, 2003
Naranus gen. nov.	83–83	(ATCC201312, MAFF425384)	*Mycol. Res.* 99: 806–808, 1995
Naranus cryptomeriae	83–83	(ATCC201312, MAFF425384)	*Mycol. Res.* 99: 806–808, 1995
Nectria asakawaensis	84–133	(MAFF240309)	*Trans. Mycol. Soc. Jpn.* 31: 228–229, 1990
Nectria hachijoensis	70–1336	(MAFF425221, NBRC32545)	*Trans. Mycol. Soc. Jpn.* 31: 228–229, 1990
Nectria fragariae	73–178		*Trans. Mycol. Soc. Jpn.* 31: 229–230, 1990
Nodulisporium melonis	75–355	(ATCC200606, MAFF425439)	*Ann. Phytopathol. Soc. Jpn.* 61: 330–331, 1995
Oedocephalum nayoroense	81–498	(ATCC201315, MAFF425312, NBRC32546)	*Mycologia* 83: 524–525, 1991
Ordus chiayiensis	98–7	(ATCC204161, MAFF237777)	*Mycoscience* 40: 383, 1999
Papulaspora nishigaharanus	73–1071	(ATCC200772, MAFF425313, NBRC32547)	*Mycologia* 83: 52, 1991
Penicillifer fragariae	73–178	Anamorph of *N. fragariae*	*Trans. Mycol. Soc. Jpn.* 31: 230, 1990
Pyrenochaeta gentianicola	75–40	(ATCC200774, MAFF455520)	*Mycoscience* 36: 443, 1995
Pyrenochaeta gentianicola	76–501	(ATCC204319, MAFF425519)	*Mycoscience* 36: 443, 1995

(Continued)

TABLE 8.2 (*CONTINUED*)
Fungi Isolated by TW and Registered as New Fungi

New Fungus	TW	Holotype	Reference
Pyrenochaeta globosa	84–523	(ATCC201414, MAFF425154, NBRC32549)	*Trans. Mycol. Soc. Jpn.* 33: 21, 1992
Sarcopodium araliae	92–53	(ATCC201313)	*Mycologia* 85: 520, 1992
Sordaria nodulifera	85–72	(MAFF425168, NBRC32551)	*Trans. Mycol. Soc. Jpn.* 30: 399–400, 1989
Sordaria tamaensis	85–89	(MAFF425169), NBRC32552)	*Trans. Mycol. Soc. Jpn.* 30: 397–398, 1989
Stachybotryna hachijoensis	70–1336	anamorph of *N. hachijoensis*	*Trans. Mycol. Soc. Jpn.* 31: 228–229, 1990
Taeniolella phialosperma	73–466	(MAFF425450)	*Mycologia* 84: 478–481, 1992
Taeniolella phialosperma	70–1070	(MAFF425240)	*Mycologia* 84: 478, 1992
Trinacrium iridis	82–567	(ATCC200608)	*Mycologia* 84: 794, 1992
Verticillium hahajimaense	00–65	(MAFF238172, CBS123923)	*Mycoscience* 42: 594, 2001
Verticillium sphaerosporum var. *bispora*	76–500		*Ann. Phytopathol. Soc. Jpn.* 46: 600, 1980

8.21 DRIED SPECIMENS OF 13 NEW FUNGI NAMED BY TW

A. *Acremonium macroclavatum* (isolate 00–50)

B. *Cylindrocarpon boninense* (isolate 00–299)

C. *Cylindrocladium tenue* (isolate 92–246)

D. *Dactylella chichisimensis* (isolate 00–315)

E. *Myrothecium dimorphum* (isolate 01–250)

F. *Nectaria asakawaensis* (isolate 84–133)

G. *Nodulisporium melonis* (isolate 75–356)

H. *Oedocephalum nayoroense* (isolate 81–49)

I. *Papulaspora nishigaharanus* (isolate 73–1071)

J. *Sordaria tamaensis* (isolate 85–89)

K. *Taeniolella phialosperma* (isolate 70–1070)

L. *T. phialosperma* (isolate 73–466)

M. *Verticillium hahajimaense* (isolate 00–65)

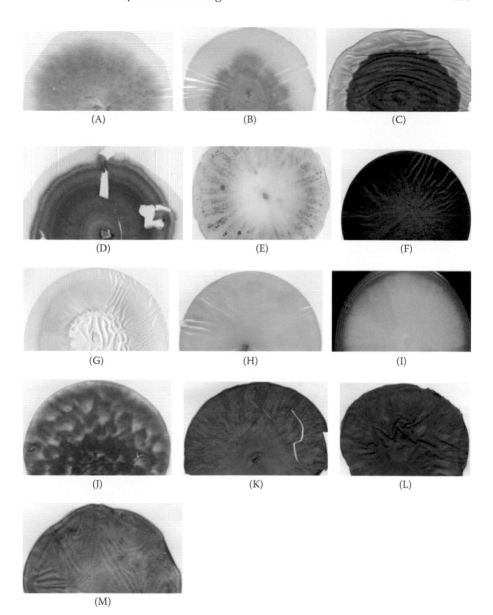

9 Soil Fungi Used for Biodegradation

9.1   WOOD DEGRADED BY TWO FUNGI SELECTED

Wood was more effectively degraded in dual cultures of *Pycnoporus coccineus* with *Flammulina velutipes, Pleurotus ostreatus,* or *Polyporus arcularius* than in their single cultures based on weight losses of chopsticks. The similar effects were attained in dual cultures of the selected Basidiomycetes and various soil fungi.

A. Dual cultures of *F. velutipes* (isolate 86–1) with either *Verticillium* sp. (center, isolate 98–53) or *Gonytrichum chlamydosporium* (right, isolate 98–64 (MAFF238006)) in relation to uninoculated control (left) together with chopsticks in plastic bottles.

B. Chopsticks in PDA as uninoculated as a control (left) and inoculated with *P. coccineus* (isolate 84–117) in plastic bottles (right).

C. Chopstick degradation due to single cultures of *P. coccineus* (middle left, isolate 84–117) and *F. velutipes* (bottom left, isolate 98–2) and a dual culture of *P. coccineus* and *F. velutipes* (right, isolate 98–2) in relation to uninoculated control (upper left).

Note the partially degraded pair of chopsticks (right) and the bleached letters on the chopsticks in the middle and bottom left.

D. Chopstick degradation due to single culture of *P. coccineus* (middle left, isolate 84–117), and unknown basidios (bottom left, isolate 321) and dual culture of *P. cocineus* and unknown basidios (right, isolate 322). Note conspicuous degradation in the dual culture.

(Watanabe et al., 2003)

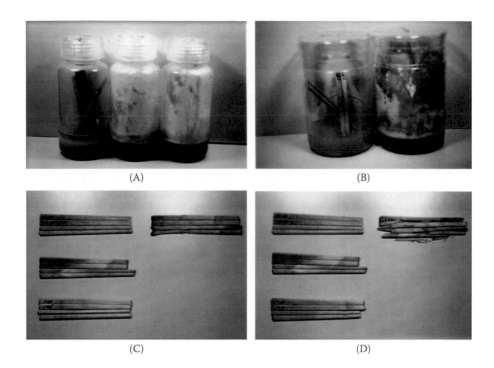

(A)

(B)

(C)

(D)

9.2 WOOD DEGRADED BY SELECTED FUNGI

A. Weight losses of a pair of chopsticks in single cultures of *P. coccineus* (isolate 84–117), *P. arcularius* (isolate 98–8), and three isolates each of *F. velutipes* (isolates 86–1, 98–1, and 98–2) and *P. ostreatus* (isolates 98–3, 98–5, and 98–6) were 25.7%, 16.6%, and 4%–10.5%, respectively, after 50-day incubation. In dual cultures of two selected fungi, *P. coccineus* (isolate 84–117) paired with either *F. velutipes* (isolates 98–1, 98–2) or *P. ostreatus* (isolate 98–5), weight losses ranged from 37%–40%.

(Watanabe et al., 2003)

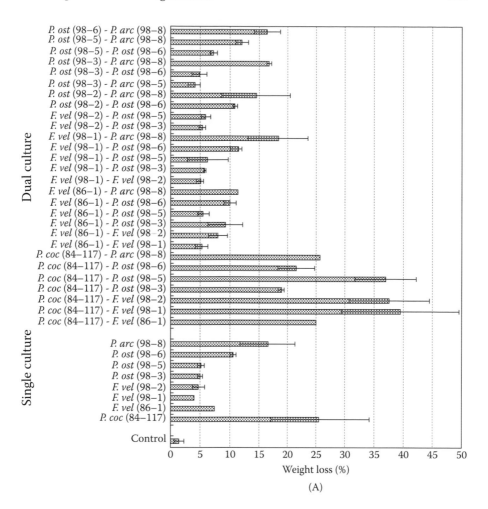

(A)

9.3 ASSAY OF CELLULOSE-DEGRADING FUNGI

Cellulose-degrading fungi were evaluated for their abilities in Difco potato dex-trose broth or Difco malt extract broth cultures with cellulose substrates in plastic Petri dishes. As cellulose substrates, filter papers (Whatman no. 3), absorbent cotton (one cut cotton, about 0.36 g), and processed cellulose powder were tested and the results were presented partially. The assay was conducted by comparing the dry weights between the uninoculated control culture weights with those of the cultures inoculated with test fungi after 30-day incubation. The dry weights were obtained in drying the cultures at 50°C for at least two days. Soil fungi were tested using at least 25 isolates. Among them, *Peniophora* sp. (isolate 06–13) and *Phlebia* sp. (isolate 99–335) were found to be the most promising as degraders. This assay tech-nique is simple, inexpensive, reproducible, and accurate as the primary screening method.

(Watanabe et al., 2012)

9.4 CELLULOSE-DEGRADING FUNGI

A–B. Disintegration of filter paper in the PD broth with filter in uninoculated control (top left), inoculated with *Gliocephalotrichum simplex* (top right, isolate 06–9 (MAFF240319)) with unknown basidios 06–13 (lower left), and with *Coprinopsis cinerea* (lower right, isolate 06–150 (MAFF240343)) in culture at 25°C for 10 days (**A**) and after drying the cultures at 55°C over two days (**B**).

C. Disintegration of cotton (two cut pieces per dish) in uninoculated control (left) and inocu-lated with *Pycnoporus coccineus* (isolate 84–117).

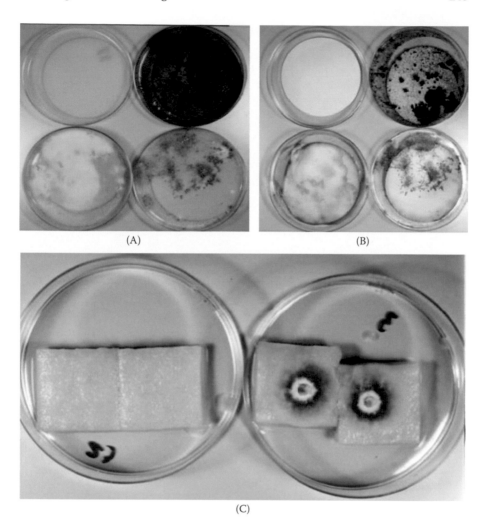

(A)

(B)

(C)

9.5 ASSAY OF FILTER PAPER-DEGRADING FUNGI

A. Disintegration of filter papers (one sheet, 0.68 g per dish) in uninoculated control and inoculated with four fungus isolates: *Gliocephalotrichum simplex* (isolate 06–9 (MAFF240319)), *Peniophora* sp. (isolate 06–13), unknown basidios (isolate 06–110 (MAFF240344)), and *Coprinopsis cinerea* (isolate 06–150 (MAFF240343)).

Peniophora sp. (isolate 06–13) was the strongest degrader of cellulose among test fungi.

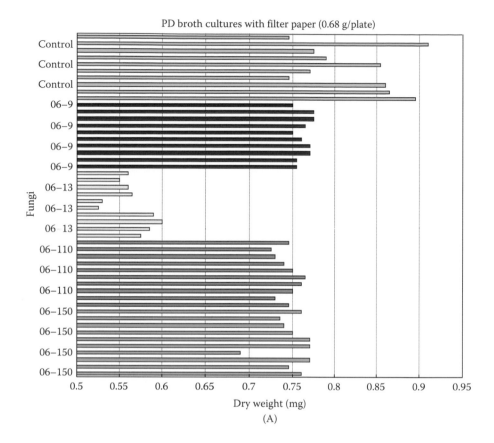

(A)

9.6   ASSAY OF COTTON- AND CELLULOSE-DEGRADING FUNGI

B. Disintegration of absorbent cotton (1.24 g/plate) in the PD broth in uninoculated control as compared with *Phlebia* sp. (isolate 99–335), *Pycnoporus coccineus* (isolate 84–117), and *Peniophora* sp. (isolate 06–13) in culture at 25°C for ten days and after drying the cultures at 55°C over two days.

C. Disintegration of cellulose powder (1 g/plate) in the PD broth in uninoculated control as compared with *Phlebia* sp. (isolate 99–335), *Pycnoporus coccineus* (isolate 84–117), *Coprinus cinerea* (isolate 06–150), and *Peniophora* sp. (isolate 06–13) in culture at 25°C for ten days and after drying the cultures at 55°C over two days.

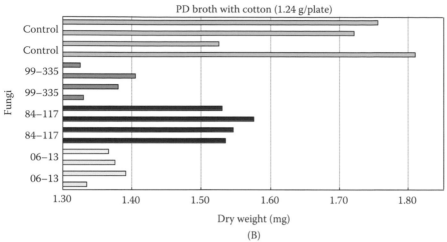

(B)

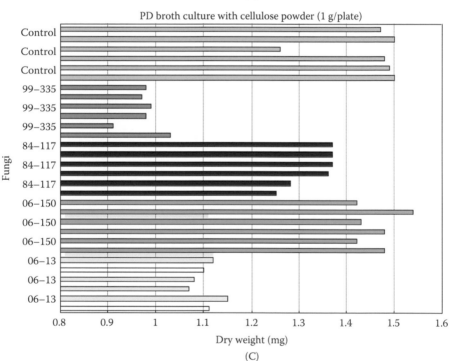

(C)

9.7   *PENIOPHORA* SP. 06–13, ONE TYPE OF CELLULOSE-DEGRADING FUNGI

D. PDA colonies of *Peniophora* sp. (isolate 06–13) in 9-cm plastic Petri dish at 25°C for 12 days.

E. Hyphae are aggregated, bundled, and branched irregularly.

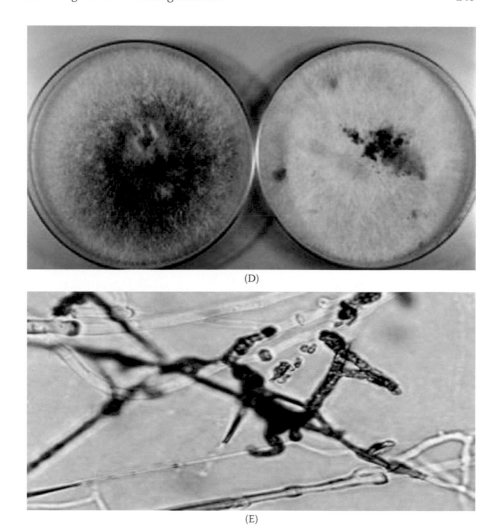

(D)

(E)

10 Field Trips

10.1   SWISS ALPS, DAVOS

The Post-Congress Field Trip of IMC 4 (Fourth International Mycological Congress, Regensburg, Germany) was held in Davos, the Swiss Alps, for nine days (September 3–11, 1989). On the occasion, soil samples were collected to analyze *Pythium* there. Some of the results were presented together with coworkers of ETH-Zentrum, Switzerland (Watanabe et al., 1998).

Temperature responses of the Swiss Alps *Pythium* isolates were also studied on that occasion (see Section 3.25, pages 70–71).

Contents of the field trip are partially presented in the following photos.

The Post-Congress Field Trip of IMC 4

A. Walking together to the destination.

B. The station house in the mountain decorated with potted flowers in the planters.

C. Tops of the Swiss Alps covered with snow, and a group walking to the destination.

D. Prof. Muller's lecture, and attendees gathering around him.

10.2   SWISS ALPS, DAVOS (CONTINUED)

The Post-Congress Field Trip of IMC 4

A. Collapsed forest trees on the mountanous area.

B. Large ant nest.

C. Discussion between Prof. Muller and one participant.

D. *Boletus* sp. distributed all over the ground.

E. *Armillaria* sp. found along the way.

10.3 FIELD TRIP TO THE OGASAWARA ISLANDS

The Ogasawara (Bonin) Islands (20°25'–24°27'N, 136°05'–153°59'E) are ocean islands located 1000 kilometers southeast of central Japan.

With Chichijima (2395 ha) and Hahajima (2080 ha) as the two main islands, the Ogasawara Islands are isolated from the main islands of Japan and have no easy access, and they are known for the biodiversity of the indigenous organisms and their evolutionary development (Ito, 1994; Ono, 1994). However, only a few works have been conducted on the soil fungi of the islands.

In my visit to the Ogasawara Islands on two occasions (January 2000 and January 2001), soil samples were collected to survey the fungal flora. Among a total of 370 isolates studied, more than 81 fungus species belonging to 47 genera were identified. Five fungi listed below were new species from the islands soil, and their species names were coined based on the Ogasawara Islands. In addition, several noteworthy fungi were isolated and identified in these two trips. Some of them were new in Japanese fungal flora.

Five new fungal species from the Ogasawara Islands, and their references, are as follows:

> *Acremonium macroclavatum* (isolate 00–50 (CBS123922, MAFF238162)), characterized by large clavate guttulate conidia (*Mycoscience* 42:591–593, 2001).
> *Cylindrocarpon boninense* (isolate 00–62 (MAFF238163)), characterized by 3–7-septate clavate macroconidia, and terminal or intercalary chlamydospores, rarely produced unicellular clavate microconidia (*Mycoscience* 42:593, 2001).
> *Dactylella chichisimensis* (isolate 00–315 (MAFF238165)), characterized by single terminal multiseptate clavate or ellipsoidal conidia at the apex of simple conidiophores, and chlamydospores and sclerotia (*Mycoscience* 42: 633–635, 2001).
> *Myrothecium dimorphum* (isolate 01–250 (MAFF23829)), characterized by dark green sporodochia composed of conidiophores with verticillate phialides and ovate and ellipsoidal, often curved conidia on their apexes mixed with erect, straight setae, and seta-like conidiophores with terminal polytomous structures composed of 2–8 digitate polyphialides bearing single globose conidia at each apex (*Mycoscience* 44: 284, 2003).
> *Verticillium hahajimaense* (isolate 00–65 (CBS123923, MAFF238172)), characterized by conidial heads bearing cylindrical conidia, and catenulate chlamydospores (*Mycoscience* 42: 594, 2001).

(Watanabe et al., 2001, 2003)

10.4 *ACREMONIUM, CYLINDROCARPON,* AND *VERTICILLIUM*

Acremonium macroclavatum (isolate 00–50 (MAFF238162))

A. Conidiophores and masses of conidia.

B. Conidia, and smooth and crustose hyphae.

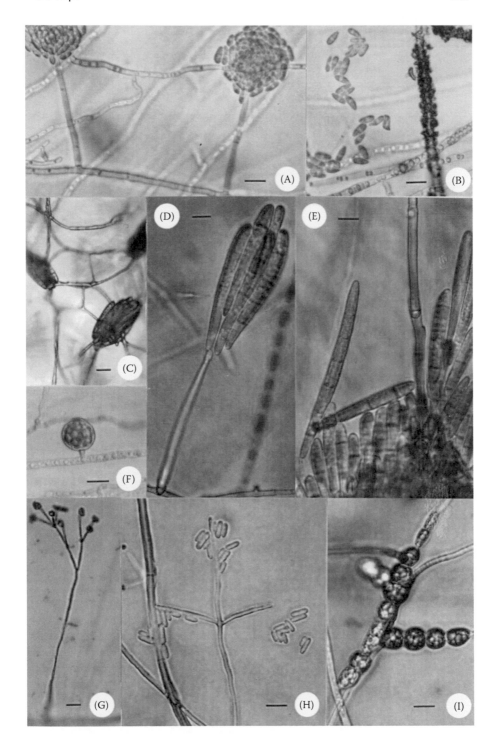

(previous page)

Cylindrocarpon boninense (isolate 00–62 (MAFF238163))

C. Habit showing aggregated sporulation.

D. Conidiophores and masses of macroconidia.

E. Aggregates of macroconidia.

F. Chlamydospore.

Verticillium hahajimaense (isolate 00–65 (MAFF238172))

G. Sporulation habit.

H. Phialide and conidia.

I. Chlamydospores in chains.

10.5   *DACTYLELLA CHICHISIMENSIS*

Dactylella chichisimensis (isolate 00–315 (MAFF238165))

A. Habit showing sporulation.

B–C. Conidiophores and conidia.

D. Conidium and chlamydospores in chains.

E. Sclerotium.

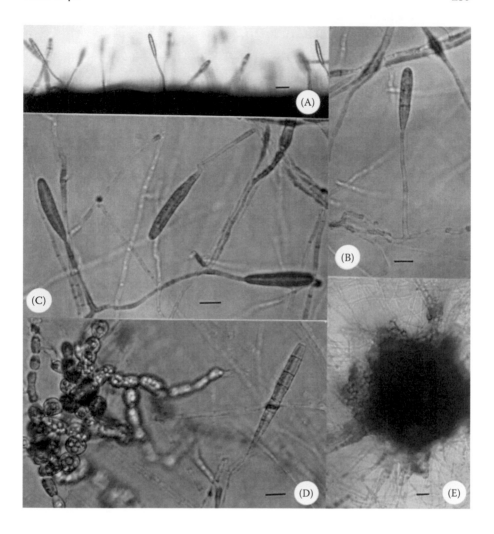

10.6   *MYROTHECIUM DIMORPHUM*

Myrothecium dimorphum (isolate 01–250 (MAFF23829))

A–B. Habit showing sporodochia with extruded several seta-like conidiophores and a seta (**A**, right).

C. Parts of sporodochium and *Gliocladium*-like setae together with an apical part of the seta-like conidiophore lying over the sporodochium.

D–E. Sporodochial conidiophores, phialides, and conidia.

F. Two seta-like conidiophores, sporodochial conidia, and two undetached globose conidia at the phialides from terminal polytomous structures with 4–8 digitate polyphialides under different focus.

G–H. One simple seta (**H**), immature (**H**) and mature seta-like conidiophores (**G**), and sporodochial conidia.

I. One simple seta, part of sporodochial conidiophores and conidia.

10.7 NOTEWORTHY FUNGI OBSERVED IN THE OGASAWARA ISLANDS SOIL

Among several noteworthy fungi isolated from soil, four fungi are illustrated: *Monacrosporium sclerohyphum* (isolate 01–183), *Neta quadriguttata* (isolate 01–481), *Pestalotia* sp. (isolate 01–164) and *Wisneriomyces javanicus* (isolate 00–273 (MAFF238168)).

A. *Monacrosporium sclerohyphum* (isolate 01–183).

B. *Neta quadriguttata* (isolate 01–481).

C. *Pestalotia* sp. (isolate 01–164).

D. *Wiesneriomyces javanicus* (isolate 00–273).

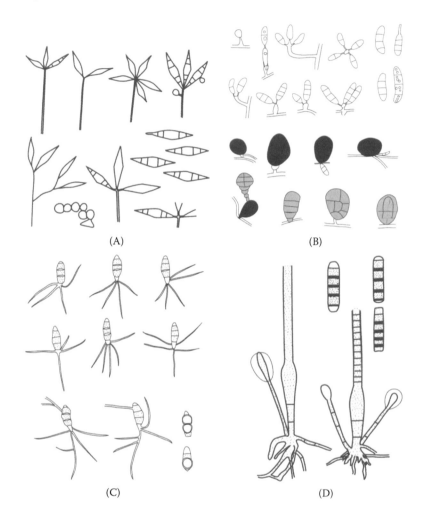

(A)

(B)

(C)

(D)

Afterword

Finally, the time has come to conclude my work on this book. Although the book's writing was based almost entirely on my own research over fifty years (1960s–2017) with a focus on soilborne diseases and causal fungi in relation to the environments, its contents must be too common, too old, or too narrow. This is because all the fungi have been studied using traditional methodologies, which predated the advent of molecular phylogenetic data. Even so, it is my hope that the accumulated evidence may influence others (and myself) to some extent. In relation to my previous work, titled *Pictorial Atlas of Soil and Seed Fungi* (3rd ed. CRC Press, Boca Raton, FL, USA), the diseased plants related to these works have been neglected with consciousness, and thus those parts have been the focus of this book. After CRC Press accepted my suggestion to write this book, I started to collect my old data and pictures, but some of them have been included repeatedly and partially in my previous works.

Although the present works have succeeded for me, the contributions and help of my co-authors have always been thoughtful through all my works.

The dry specimens have been made and preserved whenever fungi were cultured, and these specimens have always helped my works later. In addition, records of the works, including pictures and samples taken, have been preserved as much as possible.

Some of these works have been presented orally at the meetings for professional people, but some of them were not published in the journals. However, I feel very lucky to have been able to publish some of these data on this occasion.

Tsuneo Watanabe
September 16, 2017

References

Araki, T. 1967. Soil condition and the violet and white root rot diseases of fruit trees. *Bulletin of the National Institute of Agricultural Sciences, Series C.* 21, 1–109 (In Japanese).

D'ercole, N. and Canova, A. 1974. Alcuni problemi fitosanitari della fragola. *Inf. Tore fitopatol.* 9: 5–13.

Drechsler, C. 1927. Two water molds causing tomato rootlet injury. *J. Agric. Res.* 38: 287–296.

Drechsler, C. 1929. The beet water mold and several related root parasites. *J. Agric. Res.* 38: 309–361.

Drechsler, C. 1931. A crown-rot of hollyhocks caused by *Phytophthora megasperma* n.sp. *J. Wash. Acad. Sci.* 21: 513–526.

Farr, D. F., Bills, G. F., Chamuris, G. P., and Rossman, A. Y. 1989. *Fungi on Plants and Plant Products in the United States.* St. Paul, MN: APS Press, 1252 pp.

Gomez, A. L., Pires-Zottarelli, C. L. A., Rocha, M., and Milanez, A. I. 2003. Saprolegniaceae de áreas de cerrado do estado de São Paulo, SP. *Hoehnea* 30: 95–110.

Hildebrand, A. A. 1934. Recent observations on strawberry root rot in the Niagara Peninsula. *Can. J. Res.* 11: 18–31.

Ito, H. 1994. *Shima no Shokubutsushi (Plant Histories in the islands).* Tokyo, Japan: Kodansha, 246 pp. (In Japanese).

Jong, S. C., Dugan, F., and Edwards, M. J. 1996. *ATCC Filamentous Fungi,* 19th ed. Rockville, MD: American Type Culture Collection.

Jong, S. C., and Gantt, M. J. (eds.) 1987. *American Type Culture Collection Catalogue of Fungi/Yeasts,* 17th ed. Rockville, MD: American Type Culture Collection.

Kirk, P. M., Cannon, P. F., Minter, D. W., and Stalpers, J. A. 2008. *Dictionary of Fungi,* 10th ed. Wallingford, UK: CAB International.

Kurata, H. 1960. Studies on fungal diseases of soybean in Japan. *Bull. Natl. Inst. Agric Sci. Ser. C.* 12: 1–154 (In Japanese).

Matsuda, A., Moya, J. D., González, J. L., and Watanabe, T. 1998. Root rots of black pepper caused by *Pythium splendens* in the Dominican Republic. *Ann. Phytopath. Soc. Jpn.* 64: 303–306.

Meurs, A. 1934. Parasitic stemburn of Deli tobacco. *Phytopath. Z.* 7: 169–185.

Nagai, Y., Fukami, M., Murata, A., and Watanabe, T. 1986. Brown-blotted root rot of carrots in Japan. (1) Occurrence, symptoms, and isolation. *Ann. Phytopath. Soc. Jpn.* 52: 278–286.

Nagai, Y., Takeuchi, T., and Watanabe, T. 1978. A stem blight of rose caused by *Phytophthora megasperma. Phytopathology* 68: 684–688.

Nagai, Y., Takeuchi, T., and Watanabe, T. 1988. Root rot of dasheen caused by *Pythium myriotylum* Drechsler. *Ann. Phytopath. Soc. Jpn.* 54: 529–532 (In Japanese).

NBRC Catalogue of Biological Resources. *Microorganisms, Genomic DNA Clones, and cDNA Clones.* First edition 2005 Biological Resource and Center (NBRC) Department of Biotechnoloy, National Institute of Technology and Evaluation, Kazusa, Chiba, Japan.

Nemec, S. 1970. *Pythium sylvaticum*-pathogenic on strawberry roots. *Plant Disease Reporter* 54, 416–418.

Ono, M. 1994. *Kotou no Seibutsutachi (Life on the isolated islands-The Galapagos Islands and Bonin (Ogasawara) islands).* Tokyo, Japan: Iwanami Publishers (In Japanese).

Onogi, S., Uematsu, S., and Watanabe, T. 1984. Root rot of musk melon. *Plant Protection* 38: 41–44 (In Japanese).

Phytopathology Society of Japan. 2000. *Common Names of Plant Diseases in Japan,* 1st ed. Japan Plant Protection Association. 858 pp.

Pires-Zottarelli, C. L. A. 2011. Personal communication.

Plaats-Niterink, A. J. Van Der, 1981. Monograph of the genus *Pythium*. *Stud Mycol*. 21: 1–242.

Pratt, R. G. and Mitchell, J. E. 1973. A new species of *Pythium* from Wisconsin and Florida isolated from carrots. *Can. J. Bot.* 51: 333–339.

Sato, M., Watanabe, T., Furuki, I., and Morita, H. 1995. Root rot of melon caused by *Nodulisporium melonis* in Japan 1. Occurrence, symptoms, isolation and pathogenicity. *Ann. Phytopathol. Soc. Jpn.* 61: 325–329.

Shirai, M. 1906. *A List of Japanese Fungi Hitherto Known*. Yokendo: Tokyo, Japan.

Stanghellini, M. E, White, J. G., Tomlinson, J. A., and Clay, C. 1988. Root rot of hydroponically grown cucumbers caused by zoospore-producing isolates of *Pythium intermedium*. *Plant Dis.* 72: 358–359.

Strong, F. C. and Strong, M. C. 1931. Investigations on the black root of strawberries. *Phytopathology* 21: 1041–1060.

Tiffany, L. H. and Gilman, J. C. 1954. Species of *Colletotrichum* from legumes. *Mycologia* 46: 52–75.

Truscott, J. H. L. 1934. Fungus root rots of the strawberry. *Can. J. Res.* 11: 1–17.

Uematsu, S., Onogi S., and Watanabe, T. 1985. Pathogenicity of *Monosporascus cannonballus* Pollack and Uecker in relation to melon root rot in Japan. *Ann. Phytopath. Soc. Jpn.* 51: 272–276 (In Japanese).

Watanabe, T. 1972. *Macrophomina phaseoli* found in commercial kidney bean seed and in soil, and pathogenicity to kidney bean seedlings. *Ann. Phytopath. Soc. Jpn.* 38: 100–105.

Watanabe, T. 1973. Survivability of *Macrophomina phaseoli* (Maubl.) Ashby in naturally-infested soils and longevity of the sclerotia formed *in vitro*. *Ann. Phytopath. Soc. Jpn.* 39: 333–337.

Watanabe, T. 1977. Pathogenicity of *Pythium myriotylum* isolated from strawberry roots in Japan. *Ann. Phytopath. Soc. Jpn.* 43: 306–309.

Watanabe, T. 1978. Temperature-growth relations of various *Pythium* species from strawberry and sugar cane. *Trans. Mycol. Soc. Jpn.* 19: 363–372.

Watanabe, T. 1979. *Monosporascus cannonballus*, an ascomycete from wilted melon root undescribed in Japan. *Trans. Mycol. Soc. Jpn.* 20: 312–316.

Watanabe, T. 1981. Detection of *Pythium deliense* in the Ryukyu Islands and its ecological implication. *Ann. Phytopath. Soc. Jpn.* 47: 562–565.

Watanabe, T. 1983a. Black dot root rot (provisional name) of melon, Kongetsu no Nouyaku (Agrochemicals of this month) 27: 3–8 (In Japanese).

Watanabe, T. 1983b. Distribution of *Pythium aphanidermatum* in Japan: Its significance. *Trans. Mycol. Soc. Jpn.* 24: 15–23.

Watanabe, T. 1983c. Formation and deciduousness of *Pythium intermedium*. *Trans. Mycol. Soc. Jpn.* 24: 25–33.

Watanabe, T. 1984a. Damping-off diseases of vegetables, diagnosis and control. *Trans. Mycol. Soc. Jpn.* 19: 40–44 (In Japanese).

Watanabe, T. 1984b. Detection and quantitative estimations of *Pythium aphanidermatum* in soil with cucumber seeds as a baiting substrate. *Plant Dis.* 68: 697–698.

Watanabe, T. 1985. *Pythium* species found in the piedmont natural forest at Mt. Fuji. *Trans. Mycol. Soc. Jpn.* 26: 41–45.

Watanabe, T. 1986. Rhizomorph production in *Armillaria mellia* in vitro stimulated by *Macrophoma* sp. and several fungi. *Trans. Mycol. Soc.* 27: 235–245.

Watanabe, T. 1987. *Plectospira myriandra*, a rediscovered water mold in Japanese soil. *Mycologia* 79: 77–81.

Watanabe, T. 1988a. Kinds, distribution, and pathogenicity of *Pythium* species isolated from soils of various habitats in Japan. International *Pythium* Group, at the 5th International Congress of Plant Pathology and the 1st International *Pythium* workshop, *Kyoto*, Japan, 1988. pp. 3–5.

Watanabe, T. 1988b. *Pythium carolinianum* and *P. periplocum* associated with cherry seeds at the cherry tree preservation forest at Asakawa. International *Pythium* Group, at the 5th International Congress of Plant Pathology and the 1st International *Pythium* workshop, Kyoto, Japan, 1988. pp. 49–50.

Watanabe, T. 1989. Three species of *Sordaria*, and *Eudarluca biconica* from cherry seeds. *Trans. Mycol. Soc. Jpn.* 30: 395–400.

Watanabe, T. 1990. Three new *Nectria* from Japan. *Trans. Mycol. Soc.* 31: 227–236.

Watanabe, T. 1992a. A new species of *Pyrenochaeta* from Japanese black pine seeds. *Trans. Mycol. Soc.* 31: 21–24.

Watanabe, T. 1992b. Sporulation of *Dematophora necatrix* in vitro, and its pathogenicity. *Ann. Phytopathol. Soc. Jpn.* 58: 65–71.

Watanabe, T. 1992c. *Taeniolella phialosperma* sp. nov. from Japan. *Mycologia* 84: 478–483.

Watanabe, T. 1994a. *Cylindrocladium tenue* comb. nov. and two *Cylindrocladium* species isolated from diseased seedlings of *Phellodendron amurense* in Japan. *Mycologia* 86: 151–156.

Watanabe, T. 1994b. Two new species of homothallic *Mucor* in Japan. *Mycologia* 86: 691–695.

Watanabe, T. 1997. Stimulation of perithecium and ascospore production in *Sordaria fimicola* by *Armillaria* and various fungal species. *Mycol. Res.* 101: 1190–1194.

Watanabe, T. 1998. *Dictionary of Soilborne Plant Diseases.* Tokyo: Asakura Publishing Company, 272 pp. (In Japanese).

Watanabe, T. 2000. *Verticillium balanoides*, a nematode endoparasite associated with pine needles of collapsing Japanese red pine trees in Tsukuba. *Mycoscience* 41: 283–285.

Watanabe, T. 2010. *Pictorial Atlas of Soil and Seed Fungi*, 3rd ed. Boca Raton, FL: CRC Press.

Watanabe, T., Hagiwara, S., and Narita, I. 1995. Decline of *Phellodendron amurense* in Tokyo: Associated fungi and pathogenicity of associated *Cylindrocladium* spp. *Plant Dis.* 79: 1161–1164.

Watanabe, T. and Hashimoto, K. 1978. Recovery of *Gloeocercospora sorghi* from sorghum seed and soil, and its significance in transmission. *Ann. Phytopath. Soc. Jpn.* 44: 633–640.

Watanabe, T., Hashimoto, K., and Sato, M. 1977. *Pythium* species associated with strawberry roots in Japan and their role in the strawberry stunt disease. *Phytopathology* 67: 1324–1332.

Watanabe, T. and Imamura, S. 1995. Pink root rot, a revised name of brown root rot of gentian, and the causal fungi, *Pyrenochaeta gentianicola* sp. nov. and *P. terrestris* in Japan. *Mycoscience* 36: 439–445.

Watanabe, T. and Inoue, S. 1980. Root fungus floras in relation to growth of strawberry plants in Pasteurized soil in the field. *Ann. Phytopath. Soc. Jpn.* 46: 471–479.

Watanabe, T., Kanno, M., Tagawa, M., Tamaki, H., and Kamagata, Y. 2012. Primary simple assays of cellulose-degrading fungi. *Mycoscience* 53: 45–48.

Watanabe, T., Koyama, O., Kaneko, H., and Nakamura, K. 2003a. Soil fungi in the turnip field soil severely infested with the pathogen of yellows, and biocontrol trials of the disease. *Jpn. J. Phytopathol.* 69: 288–289 (Abstr. in Japanese).

Watanabe, T., Nagai, Y., and Fukami, M. 1986. Brown-blotted root rot of carrots in Japan. (2) Culture and identification. *Ann. Phytopath. Soc. Jpn.* 52: 287–291.

Watanabe, T. and Nakamura, K. 2008. Diversities of rice seed-associated fungi and the significance. *Jpn. J. Phytopathol.* 74: 184–185 (Abstr. in Japanese).

Watanabe, T., and Narita, I. 1994. *Irpicomyces cornicola* sp. nov. from *Cornus florida* in Japan. *Mycoscience* 35: 105–108.

Watanabe, T., Onogi, S., Uematsu, S., Izumi, S., Kodama, K. and Tanaka, T. 1983. Relation of black dot root rot (provisional name) of melon occurred in various parts of Japan, and *Monosporascus cannonballus. Ann. Phytopath. Soc. Jpn.* 49: 127 (Abstr. in Japanese).

Watanabe, T., Onogi, S., Uematsu, S., and Tsuchiya, Y. 1983. Root rot of musk melon caused by *Pythium splendens. Ann. Phytopath. Soc. Jpn.* 49: 127 (Abstr. in Japanese).

Watanabe, T., Ozawa, M., and Sakai, R. 1973. A new disease of Chinese cabbage caused by *Verticillium albo-atrum* and some factors related to the incidence of the disease. *Ann. Phytopath. Soc. Jpn.* 39: 344–350.

Watanabe, T. and Sato, M. 1995. Root rot of melon caused by *Nodulisporium melonis* in Japan. 2. Identification. *Ann. Phytopath. Soc. Jpn.* 61: 330–333.

Watanabe, T., Sieber, T. N., and Holdenrieder, O. 1998. *Pythium* in the Swiss Alps. *Mycologia Helvetica* 10: 3–13.

Watanabe, T., Tagawa, M., Tamaki, H., and Hanada, S. 2011. *Coprinopsis cinerea* from rice husks forming sclerotia in agar culture. *Mycoscience* 52: 152–156.

Watanabe, T., Tzean, S. S., and Leu, L. S. 1974. Fungi isolated from the underground parts of sugar cane in relation to the poor ratooning in Taiwan. *Trans. Mycol. Soc. Jpn.* 15: 30–41.

Watanabe, T., Uematsu, S., and Hayashi, K. 1987. Fungal isolates from seeds of two cherry species collected at the cherry tree preservation forest at Asakawa. *Trans. Mycol. Soc. Jpn.* 28: 475–481.

Watanabe, T. and Umehara, Y. 1977. The perfect state of the causal fungus of bakanae disease of rice plants recollected at Toyama. *Trans. Mycol. Soc. Jpn.* 18: 136–142.

Watanabe, T., Watanabe, Y., and Fukatsu, T. 2002. Diversity of *Pythium* species in the Bonin (Ogasawara) Islands. *Ann. Phytopath. Soc. Jpn.* 68: 67 (Abstr. In Japanese).

Watanabe, T., Watanabe, Y., Fukatsu, T., and Kurane, R. 2000. Possibilities of *Mortierella alpina* and *M. tsukubaensis* sp. nov. as biocontrol agents against *Pythium* and *Rhizoctonia* species. *Ann. Phytopath. Soc. Jpn.* 66: 180 (Abstr. in Japanese).

Watanabe, T., Watanabe, Y., Fukatsu, T., and Kurane, R. 2001a. *Mortierella tsukubaensis* sp. nov. from Japan, with a key to the homothallic species. *Mycol. Res.* 105: 506–509.

Watanabe, T., Watanabe, Y., and Fukatsu, T. 2001b. Soil fungal floras in the Bonin (Ogasawara) Islands, Japan. *Mycoscience* 42: 499–502.

Watanabe, T., Watanabe, Y., and Nakamura, K. 2003. Biodegradation of wood in dual cultures of selected two fungi determined by chopstick method. *J. Biosci. Bioeng.* 95: 623–626.

Wilhelm, M. S., Nelson, P. E., Thomas, H. E., and Johnson, H. 1972. Pathogenicity of strawberry root rot caused by *Ceraobasidium* species. *Phytopathology* 62: 700–705.

Yoshida, S., Murakami, R., Watanabe, T., and Koyama, A. 2001. *Rhizopus* rot of mulberry-grafted saplings caused by *Rhizopus oryzae. J. Gen. Plant Pathol.* 67: 291–293.

Zenbayashi, R., Shibukawa, S., and Watanabe, T. 1985. Petiole rot of arrowhead caused by *Pythium myriotylum. Ann. Phytopath. Soc. Jpn.* 51: 482–485 (Abstr. in Japanese).

Index